**1981**

# Evaluating Built Environments:

# a behavioral approach

## by Robert W. Marans & Kent F. Spreckelmeyer

Published jointly by:

Survey Research Center
Institute for Social Research
The University of Michigan

and

Architectural Research Laboratory
College of Architecture and Urban Planning
The University of Michigan

The study described in this volume was conducted under a grant from the U.S. Department of Commerce, National Bureau of Standards (NBS), Grant Number G8 9020. (NBS has assigned the report identification number NBS-GCR-81-303 to this volume.)

*T H*
*6 025*
*. M 37*

Library of Congress Cataloging in Publication Data:

Marans, Robert W.
  Evaluating built environments.

  Bibliography: p.
  1. Buildings — Environmental engineering — Psychological aspects.  2. Office buildings — Michigan — Ann Arbor — Environmental engineering — Psychological aspects.
I. Spreckelmeyer, Kent F., 1950–   .  II. University of Michigan.  Survey Research Center.  III. Title.
TH6025.M37      725'.1      81–6709
ISBN 0-87944-272-7      AACR2


ISR Code Number 4470

Published jointly in 1981 by:
  Institute for Social Research,
  The University of Michigan, Ann Arbor, Michigan
and
  Architectural Research Laboratory,
  College of Architecture and Urban Planning,
  The University of Michigan, Ann Arbor, Michigan

6 5 4 3 2 1
Manufactured in the United States of America

Jacket and Cover Design by William Beckley, University of Michigan Publications Office

# Acknowledgements

This study has benefitted from the contributions of many people. Most are associated with The University of Michigan's College of Architecture and Urban Planning and its Architectural Research Laboratory (ARL). The college is headed by Dean Robert C. Metcalf, and Professor Jonathan King serves as the director of ARL. Professors King and Robert Johnson served as consultants to the project and made helpful comments and suggestions throughout the course of the work. Guidance on the selection and use of technical instrumentation for the measurement of the environmental data was generously provided by Professor Norman Barnett. Four graduate students in architecture who participated in an evaluation seminar taught by Professors Marans and King also participated in this study. Tim Dodson, Jeffrey Hanthorne, Roberta Oakley, and Vira Sachakul played important roles during the research design and data collection phase. Mr. Sachakul also prepared the sketches used in this report.

Several people at U-M's Institute for Social Research also participated in the work. We wish to acknowledge the assistance of Barbara Thomas, Pamela O'Connor, Diane Carbone, Laura Klem, and Betty Zebell. Ms. Zebell was instrumental in the production of this report, including the preparation of tables and the final manuscript. Our thanks to Stanley Seashore and Lee Sechrest, who read an early draft of the manuscript and offered a number of valuable suggestions, and to Linda Stafford, whose contributions as editor have greatly enhanced the readability of this volume.

Special thanks are also extended to Louis Harris and Associates for their cooperation in sharing unpublished data from their national study of office workers, and to Wystan Stevens, Ann Arbor city historian, for sharing insights into the historical development of the Ann Arbor Federal Building. Additional information about the historical background of the building was provided by Edward Buddress, Jerry Diptolla, and Don Zbylut of the Chicago regional office of the General Services Administration. We also thank Nils Strand, GSA's building manager, for his cooperation and support in the evaluation.

The Ann Arbor Federal Building was designed by the architectural firm of Tarapata, MacMahan, and Paulsen of Bloomfield Hills, Michigan. Glen Paulsen, the principal-in-charge, Larry Morris, the project architect, and Karen Ashker, the interior designer, all provided invaluable information on the planning and execution of the building's design.

The study was supported by a grant from the Center for Building Technology, National Bureau of Standards, U.S. Department of Commerce, Grant Number G8-9020. Dr. Stephen Margulis, of the Bureau's Environmental Design Research Division, served as grant monitor and provided useful guidance and assistance throughout the work.

Finally, the cooperation and insights of the federal employees who participated in the study are gratefully acknowledged. It is our hope that the material presented here will be helpful to them in maintaining and improving the quality of their work lives and contribute to more satisfying work environments for workers elsewhere.

# Contents

CONTENTS

# 1

# Introduction

In recent years, an increasing number of policy makers, environmental design practitioners, and researchers have recognized that more systematic information is necessary as a basis for environmental decision making, planning, and programming. As part of their work, many have turned to diagnostic or post-occupancy evaluations of built environments as sources of needed information. These evaluations have been designed in part to determine the extent to which the objectives of clients and designers have been fulfilled. At a time when unfulfilled objectives are costly to rectify in both economic and social terms, rational approaches to making decisions and assessing the outcomes of previous decisions are becoming increasingly important. These views are shared by a variety of organizations and professional groups operating in the building field.

The United Nations Center for Housing, Building, and Planning, for example, as part of their program for promoting social integration through housing programs, has recently published a report covering four case studies involving post-occupancy or neighborhood evaluations. Specifically, the case studies consider the extent to which physical and other attributes of neighborhoods in four countries contributed to the objective of achieving integration of their diverse population groups (United Nations, 1978).

In another context, the General Services Administration (GSA) of the U.S. government has been developing an evaluation program whereby they can learn about federal installations and work environments and use the information as a basis for programming new facilities built under their sponsorship. At the same time, the Senate Committee on Environment and Public Works has addressed the issue of research and evaluation as part of the Public Buildings Act of 1980. Section 108 of that act requires the GSA administrator "to carry out systematic research and post-occupancy evaluations" and authorizes demonstration projects "to determine and improve the effectiveness of existing and planned public buildings in providing productive, safe, healthful, economical, conveniently located, energy efficient and architecturally distinguished accommodations for federal agency offices." In recent years, the federal government has demonstrated a commitment to quality architecture through their process of selecting architects for public buildings (Architectural Record, 1978). Whether the government will make a concerted effort to follow the Section 108 directive with respect to research and evaluation remains to be determined.

During the past decade, evaluation studies on building environments ranging from new towns to health-care facilities and public and private housing have been prepared by environmental design researchers, often working in collaboration with government agencies, student groups, or practitioners (Lansing et al., 1970; Cooper, 1975; Friedman et al., 1978). Additional collaborative efforts will most likely take place in the 1980s. A recent editorial in a prominent architectural journal posited the value of user reaction studies and post-occupancy evaluations and suggested that during the next decade these activities are likely to mature as a segment of professional practice (Progressive Architecture, 1980).

Paralleling an increasing number of post-occupancy evaluations has been a growing concern that the procedures used to conduct many of these evaluations have not been systematic (Marans, 1978; Canter et al., 1980). Few attempts have been made to gather the necessary data in an orderly manner or to analyze them in such a way that the results can have both immediate and long-term applicability. Furthermore, the approaches to evaluation vary greatly, and few have been based on well-developed conceptual models. For instance, a variety of evaluations have relied on questionnaires administered to building occupants in order to determine the extent to which they use and like the building and its various attributes. Other studies have attempted to assess specific environmental conditions such as noise and light levels and the amount of space available to building occupants. Yet few evaluations have gathered both types of data and examined them with respect to one another. In part, researchers and environmental designers have agreed that these limitations have been largely a function of scarce financial support necessary for the careful design and systematic execution of building evaluations.

## Purpose and Scope of the Study

In an attempt to overcome some of these limitations, this monograph presents a systematic approach to designing and implementing evaluations of built environments. It does so by presenting a case study focusing on one particular built environment — a federal office building in Ann Arbor, Michigan.

Several factors influenced our decision to choose the Ann Arbor Federal Building for this case study. Being relatively new — it was first occupied in 1977 — it had been built under new federal guidelines calling for architectural excellence. Second, it was recognized for design excellence by the architectural profession. The building received several design awards and extensive publicity in newspapers and in the architectural press. Nonetheless, it was reputed to have problems and has been the focus of controversy within the Ann Arbor community since its downtown site was first announced in the early 1970s. Third, the building is located in close proximity to the offices of the principal researcher. Finally, choosing the Ann Arbor Federal Building for the case study offered the potential for adapting the findings and the approach used in this evaluation to other built environments, including those built under federal sponsorship.

This evaluation has been made from a single perspective — that of the building users. The major users are the federal employees who work in the building, and the residents of Ann Arbor and its surrounding communities are a second group of building users. Information about these two groups and how they interact with the building was obtained through questionnaires administered to all of the building employees and to samples of community residents, through measures of a number of specific environmental characteristics of the building, and through systematic observations of both user groups.

The self-administered questionnaire was completed by 239 federal employees — more than 90 percent of the people working in the 14 separate agencies housed in the Ann Arbor Federal Building. The questions focused on their activities, how they felt about the building as a place to work, and how they rated the building's appearance and a number of specific environmental attributes. Interviews with two groups of community residents addressed their use of the building and their feelings about its overall design. Via telephone, we contacted 113 adults from the Ann Arbor community who were selected by probability methods. In addition, we interviewed 60 building visitors at the site. Each of the interviews with members of the community lasted about 10 minutes; the questionnaires administered to building occupants were designed to be completed in about 15 minutes.

Information describing various environmental conditions was gathered from within each of the agencies and at individual work stations. Data were collected on lighting, temperature, humidity, noise, furniture and equipment arrangements, and the amount of workspace. The extent to which the attitudes and behaviors of employees were related to these environmental conditions was then examined.

The findings indicate that the Ann Arbor Federal Building is successful in at least one major respect. It has become an integral part of downtown Ann Arbor and has contributed to the attractiveness and economic vitality of the area. It is readily identifiable and is used with regularity by the public. Most of the people we questioned considered the building to be both worthy of its design awards and conveniently located. People who worked in the building, too, liked its location and were able to make extensive use of nearby shops, banks, restaurants, and other services.

However, the building has not lived up to its expectations of providing a high quality work environment for all of its occupants. One-third of the people employed in the building expressed dissatisfaction with their immediate workspace, and one-quarter gave poor ratings both to the building's appearance and to the spatial arrangement of the agency with which they were employed. Work station dissatisfaction was associated with having little privacy, no windows or windows showing unattractive views, too much noise, and uncomfortable variability in temperature. These conditions were most prevalent in the open-office settings characteristic of many of the agencies. Opinions about workspace were likely to color people's general reactions to the building and specifically to its architectural quality. Despite its favorable public image, many of the people who worked in the building considered it to be aesthetically and functionally deficient.

In part, worker dissatisfaction can be linked to the flexible spaces that were designed to accommodate changes both in government agencies and in internal agency functioning. The provision for flexibility in the building design was not supported by the day-to-day management and operation of the building.

## Users of the Study

This study was undertaken with several audiences in mind. Our findings should suggest to architects and space planners the value of examining the impact of design solutions on the people who will eventually occupy and use the structures and spaces they create. Designers should be interested in knowing if their environments actually function in the manner in which they were intended. Are the spaces supportive of the work activities, or does the environment inhibit or restrict successful completion of those activities? Are the forms and spaces satisfying to the workers and to the public? Is the building aesthetically pleasing, and was the choice of site a correct one? Questions such as these can be important to designers in organizing and carrying out future work. The material offered in this monograph suggests ways of posing these questions and seeking their answers.

Although this evaluation focuses on a single office building designed and erected under the sponsorship of the federal government, we believe our findings can be useful to architects and designers of work environments in other settings as well. We have assessed how people respond to different office arrangements and degrees of spatial separation, and we have related specific elements of the physical environment to worker satisfaction and job performance. In the final chapter, we discuss these and other findings in light of the original design objectives set forth by the architects of the Ann Arbor Federal Building and their client.

As we mentioned briefly, many of the problems with the Ann Arbor Federal Building and with its flexible, open-office arrangement can be attributed to improper facility management. Clearly, there are lessons that building managers and others responsible for office space can learn from this evaluation. One lesson suggests that flexibility in building design needs to be accompanied by a carefully developed management plan and day-to-day execution of that plan. Changes in furniture arrangement, for example, necessitate changes in the location of electrical outlets, communications systems, and lighting; these must be planned for and made with as little disruption as possible to the tasks of workers and the aesthetic quality of the space.

Within the federal government and particularly within the General Services Administration, administrators, space planners, and building managers can learn from our experiences in doing this evaluation and from our findings. Throughout the monograph, we compare environmental conditions and employee responses to those conditions for the several agencies housed in the building. These comparisons can serve to highlight differences in what people have, in what they do, and in how they feel about their jobs and their work environments.

Finally, environmental researchers can benefit from this work by reviewing and critically appraising our approach to the evaluation of a particular structure. A major, systematic effort was made to gather and analyze people's responses to environmental conditions, but we did encounter a number of methodological and theoretical problems which are discussed in Chapters 2 and 9. Many of these problems warrant further study; specifically, we feel that more work is needed on the development of techniques for measuring environmental conditions. Attention should also be given to measuring complex behavioral patterns and using them to examine relations between environmental conditions and people's subjective responses to their environment.

## Organization of the Report

This first chapter has dealt with the background of the study, its methodology, and some of the major descriptive findings. In Chapter 2, we discuss our approach to conducting this evaluation and present a conceptual model showing how we examined data on the environment and the users' responses to it. Chapter 3 outlines the history of the Ann Arbor Federal Building and describes its setting, its design, the agencies that occupy it, and the building users. The objectives of the building as described by GSA representatives and the building's architects are outlined in Chapter 4 along with a summary of the major evaluative issues addressed in subsequent chapters.

In Chapter 5, we discuss relationships between the building and its surroundings. We consider, from several perspectives, the degree to which the building has been successfully integrated into the downtown area. These deal with the attitudes of both the public and the building occupants and with the latter's use of nearby downtown facilities. Transportation and the parking situation near the building are also discussed in Chapter 5. In Chapter 6, we look at the ways in which the building is used by the public and by the people who work there and how both groups evaluate the overall building design.

Attention is directed in Chapter 7 toward the work environment within the Ann Arbor Federal Building and how the workers experience it. We first discuss the work environment at two levels — within the agencies that occupy the building and at the agencies' individual workspaces. We then examine how these levels of the work environment are viewed by the workers. In Chapter 8, we explore the issue of job performance and the extent to which it is influenced by the work environment.

In the final chapter of this monograph, we summarize the major findings and discuss our conclusions about how successful the building has been in fulfilling its intended objectives. We also discuss the changes that have taken place in the building after our evaluation but prior to the preparation of this report. Finally, a series of recommendations is outlined covering policy issues, design considerations, and future work on post-occupancy evaluations.

# 2

# An Organizational Framework for Conducting Evaluations of Built Environments

In this chapter, we present an overview of our approach in evaluating the Ann Arbor Federal Building and describe the conceptual model that guided the analysis of the data collected as part of the evaluation. Although the specific activities and the kinds of data collected and analyzed are unique to this study, the overall approach we used and our conceptual model can serve as an organizing framework for systematic evaluations of built environments in other settings.

## Overall Approach

Our evaluation of the Ann Arbor Federal Building consisted of four overlapping phases, including many activities that were performed simultaneously:
1. a preliminary exploration or reconnaissance of the building — to learn about its historical development and to identify problems and issues that could be addressed systematically;
2. a research design phase — to determine data needs and the approaches that would be used to gather data and to design and test data-collection instruments;
3. a data collection phase — to administer the questionnaires, complete the site observations, and measure environmental conditions;

4. a documentation, data analysis, and dissemination phase — to code the various types of information that had been collected, build computer files, begin the iterative process of data analysis, and report our preliminary findings.

Figure 2.1 is a chronological diagram characterizing the evaluation process for this project. Following preliminary contacts with General Services Administration (GSA) officials and authorization to conduct the evaluation, intensive work began in the late summer of 1979 and lasted for approximately 16 months.[1] The four phases covering this period are shown in the upper half of the diagram, and the specific activities included in each phase are depicted in the bottom portion of the diagram.

*Reconnaissance Phase*

This initial phase consisted of two concurrent sets of activities — a series of meetings with individuals involved in the inception, design, management, and use of the building and numerous visits to the building and its environs. The preliminary meetings with GSA officials addressed the initial plans for the building, the purposes it was intended to fulfill, and the manner in which it was operating. At the first meeting, the operations manager for the building — whose offices were 80 miles away, in Battle Creek, and who visited Ann Arbor every few weeks — gave a general overview covering these points and identified a number of problems and issues that might be considered as part of our evaluation. The operations manager also identified key personnel in the design branch of GSA who had been responsible for the building program and for supervising the design and construction. Subsequently, these individuals provided a detailed history of the development of the building, including the conceptual thinking underlying its physical design.

A second set of meetings was held with the building's architects and interior designers. They, too, contributed to our understanding of the building's history and the purposes it was intended to fulfill. In addition, they described to us the philosophy behind their decisions about spatial and functional arrangements and the manner in which these notions were translated into physical form. They also provided detailed plans, renderings, design calculations, and the user questionnaires they had employed to delineate individual and organizational spatial requirements.

A third set of meetings was held with the heads of the federal agencies occupying the building. They were questioned about the purposes and composition of their organizations, the ways their staff used the

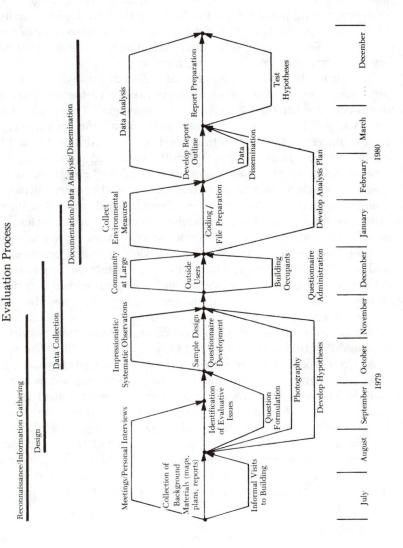

FIGURE 2.1

Evaluation Process

new facility, and their agencies' prior locations. They were also asked to convey their feelings about the building and the kinds of problems they or their staffs had experienced since moving in. As a result of these interviews, we began to recognize that a much wider range of functions was taking place within the building than we had originally envisioned. In the final chapter of this volume, we discuss some of the implications of attempting to evaluate a building containing numerous organizations with greatly varied functions.

In addition to making a number of unscheduled visits to the building, members of the research team were taken on a formal tour during our meeting with the GSA operations manager. This enabled us to meet the people who were responsible for managing each agency and to familiarize ourselves with those parts of the building that are inaccessible to the public. Subsequent visits were made to each agency to observe informally the way spaces were being used and to make photographic records of various activities, spaces, and problem situations.

Following our review of what had been learned from these informal visits, we planned another set of visits to each of six agencies during specified periods in order to obtain a concensus view on the movements of people, communication flows, office decor, spatial arrangements, and ways in which the physical environment might be supportive of or detrimental to aspects of individual jobs or organizational functioning. These impressionistic observations were made within the offices of the following agencies: the Internal Revenue Service, the Social Security Administration, Heritage Conservation and Recreation Service, the National Weather Service, the Post Office, and the military recruiters.

As a result of both the meetings with GSA representatives, the architects, and agency heads and our visits to the building, we were able to complete a preliminary listing of key issues to be examined as part of the evaluation. Members of the project staff suggested a number of research questions and hypotheses, and together we specified the kinds of data necessary to address these questions and test the hypotheses. At this time, we planned how data of an attitudinal and behavioral nature were to be obtained from various users of the building, and we specified the kinds of environmental data that would be required from measurements of conditions in the building.

*Research Design Phase*

Information was needed from two groups of building users — both from the people who worked there and from those community residents who visited the agencies occupying the building. In drafting the questionnaires to be administered to the two user groups, we primarily

wanted to address the evaluative issues; most of the questions were designed to yield data that would enable us to either describe a particular
situation or test a specific hypothesis. Other questions were proposed so
that our results could be compared with results of evaluations of comparable environments. For example, it was proposed that a number of
questions used in a recent national survey of office workers[2] should be
asked of the federal building occupants so that their responses might be
examined and compared to the responses of office workers in other settings. Two slightly different questionnaires were needed for our samples
of community residents — one for visitors to the building who were to
be interviewed at the site and another for residents throughout the
community-at-large who were to be queried by telephone.

Each of our three questionnaires was pretested by members of the research team who had been trained in questionnaire administration in
early November. A University of Michigan professor who had worked
part-time for one of the federal agencies reviewed the building occupants questionnaire and suggested a number of revisions; two of his
students who had worked in the Ann Arbor Federal Building were also
asked to respond to the questionnaire. On the basis of their responses,
another pretest questionnaire was developed and administered to
approximately 20 employees in a new federal office building in
Saginaw, Michigan; their written and verbal comments aided us in
preparing the final occupants questionnaire.[3]

Similarly, the questionnaires to be administered to the community
residents were pretested in both face-to-face interviews with people in
the lobby of the Ann Arbor Federal Building and in telephone interviews with people whose telephone numbers were not selected in our
random sample.

Because only a relatively small number of people work in the Ann
Arbor Federal Building, all building occupants who were employed in
the building during a given week — the last week in November 1979 —
were given questionnaires. Sampling procedures had to be devised for
the other two groups to be questioned as part of the study. Our residents group was chosen through a probability sample of 174 residential
telephone numbers selected from the Ann Arbor telephone directory.[4]
A quota sample of outside users of the building was designed to obtain
60 interviews during a one-week period. Members of the research team
were stationed at building entrances at specific time periods for several
days and administered the questionnaires to adults who were leaving or
entering the building.

As part of the design phase, procedures for collecting quantitative
environmental data were also prepared. These data were to describe

more fully a number of the building's physical characteristics or attributes covered in our user questionnaires. Information was to be gathered through field investigations and indirectly from the working drawings and floor plans showing furniture arrangements. The locations and layouts of the work stations were verified using the floor plans for each agency. Finally, we developed and pretested instruments for measuring and recording specific types of environmental data that were to be obtained by direct and indirect methods.

In addition to the quantitative data on users of the building and on the environments within which agency personnel worked, we sought quantitative data on the use of the building's public areas: the entrances of the building, the information desk in the main lobby, the snack bar on the second floor, and the lounge area outside the snack bar. Procedures were developed for obtaining systematic counts on the number of people using each of the four building entrances, the extent to which people sought assistance at the information desk, and the extent to which employees and the public used the snack bar and its lounge.

*Data Collection Phase*

As noted above, the attitudes of building users were measured through questionnaires administered to the occupants, on-site visitors, and Ann Arbor residents. The three questionnaires were also designed to provide some behavioral data, which were supplemented with observations made at the building. Environmental data, on the other hand, were obtained in a direct manner by visiting the building and measuring specific physical attributes or by taking measurements from the plans showing furniture arrangements in each agency.

Our initial effort at systematic data collection focused on the uses of the public areas of the building, including its entrances. Observations were made over a one-week period in late October.

Because it was known that the heaviest use of the entrances would occur during the early morning hours and late in the day when federal employees came to and left work, full hourly counts were made between 7:30 and 8:30 a.m. on two mornings (Monday and Tuesday), and between 3:30 and 4:30 p.m. on two different afternoons of the week (Wednesday and Friday). Similarly, because public use of the building was expected to be high during the lunch hour, we also completed two separate hourly counts from noon to 1:00 p.m. (on Tuesday and Friday). A sample of time periods was the basis for subsequent counts of building entrance use during the remaining daytime hours of the week. Observers counted persons at the four building entrances during each of ten half-hour periods.[5]

Systematic counts were also made of the number of people who talked with the security guard at the information desk and the number who used the lounge area outside the coffee shop/snack bar on the second floor of the building.[6]

Shortly after completing the systematic observations, we initiated observations of a more impressionistic nature in each of six key agencies to obtain a better understanding of the activities of people, assess the office arrangements and decor, and gauge the interactions between the workers and their environment.

Within each agency, observations were made by five members of the research team, who each spent 15 minutes at a preselected observation point. Thus, for each agency, observational data were obtained covering specific attributes and behaviors occurring over a 75-minute period of the work day. The team members then met to review their impressions and reach a consensus on several characteristics of the agencies. These impressionistic observations are summarized in Chapter 6.

After the questionnaires had been developed and pretested and the sampling procedures had been designed, the collection of survey data was begun. Letters had been previously distributed to the building occupants informing them of the study and its purpose and asking for their cooperation in completing the questionnaire. The letter also described our data collection procedures and the manner in which the anonymity of their responses would be guaranteed.[7]

Questionnaires were distributed to 270 building occupants on the Monday morning following Thanksgiving.[8] Sealed collection boxes had been placed at a conspicuous location within each agency. Questionnaires were picked up by members of the research team at the end of each day that week and, by Friday afternoon, a total of 239 questionnaires had been returned, representing an 88.5 percent response rate.[9]

Beginning in mid-November, telephone interviews were conducted with residents identified through the random sample of Ann Arbor telephone numbers. The 174 residential telephone numbers in the sample yielded 113 successfully completed interviews, for a response rate of 83.3 percent.[10]

Interviews taken at the building with the quota sample of outside users were conducted during the final week in November. By mid-December, all the interviewing was completed and the questionnaires covering the three groups of users had been logged in and prepared for subsequent coding.

The second major data collection effort began in December with a recording of environmental data using the working drawings and floor plans showing furniture arrangements. Data covering each work station were recorded on forms that had been pre-numbered to correspond to the questionnaires administered to the building occupants. These "in-direct" data included the square footage devoted to each work station, worker density, and distances to agency entrances, windows, and lightwells.[11]

Beginning in mid-January 1980, more direct environmental mea-sures were recorded for 265 work stations — covering noise levels, tem-perature, humidity, light levels, and glare conditions. These data were recorded on another set of pre-numbered forms.[12] Measurements for light and noise levels, temperature, and humidity were taken at two different times over a period of one month. The specific types of mea-sures taken at the building and the equipment used are described below.

*Light measurement.* A hand-held "Photo Research" model 501 digi-tal photometer was used to measure light levels at each work station. The meter was placed at the center of the employee's work surface and a foot candle reading was taken; any task lighting was switched on prior to the meter reading. For work stations not having access to nat-ural light — that is, work stations located more than 20 feet from win-dows and more than 10 feet from lightwells — only one reading was taken. At the naturally lighted work stations, two readings were taken — one on a sunny day and another on a hazy day.[13]

*Noise measurement.* A hand-held "Ivie" Model IE-10A noise meter was used to record sound levels for several zones throughout the build-ing. In most locations, a single reading was made. However, two read-ings were taken within the Post Office and in agencies with a lightwell above or below. The second Post Office reading was taken during the early morning hours when the mail was being sorted and noise levels tended to be higher. Each reading involved three measurements: (1) a decibel reading to identify the noise intensity and to assess an appro-priate level of sound across a range of frequency bands; (2) a "Noise Criteria" (NC) reading to assess noise intensities at various Hertz levels; and (3) a Hertz-level reading at which the NC was greatest.[14]

*Temperature and humidity.* As with the noise measurements, two temperature and relative humidity readings were taken within specific zones throughout the building. Readings were made with a "Bacharach" cyclometer.

*Glare.* On the basis of discussions with lighting engineers and others with experience in measuring glare, we decided to use a general rather than a detailed approach to assessing indoor levels of glare conditions; a detailed approach would have entailed more time and more equip-ment than was available to our research team. The general approach involved categorizations of each individual's seating position or orien-

TABLE 2.1

Summary of Environmental Measures Used in This Study

| Direct | Indirect |
|--------|----------|
| Temperature (Z) | Amount of workspace (W) |
| Relative humidity (Z) | Density of workspace (W) |
| Light level (W) | Type of workspace (W) |
| Noise level (Z) | Glare condition (W) |
| Style of chair (W) | Distance to window (W) |
| Use of task lighting (W) | Distance to lightwell (W) |
| Use of extension cords (W) | Distance to entrance (W) |
| Use of personal objects (W) | Distance to coffee station (W) |

Note: Z represents measures made in zones within agencies. These measures were then assigned to individual work stations within the respective zone. W represents measures covering individual work stations.

tation vis-à-vis natural lighting. The categorizations were made using floor plans and were later verified at the time the direct measures were being taken.

In addition to making the above measurements of ambient environmental conditions, other characteristics of individual work stations were identified and recorded while the research team was at the building. Included here were two behavior measures — the nature of the tasks being performed at each work station and the extent to which each work station was personalized. Similarly, data were also gathered for other conditions that might have some influence on worker satisfaction: the type of chair at the work station, the presence or absence of task lighting, and the presence or absence of electrical and telephone extension cords. A summary of the environmental data collected both directly and indirectly is shown in Table 2.1.

Two issues related to the collection of data should be noted here. The first deals with the dynamic nature of the work stations and the people who occupy them. Both people and furniture arrangements were constantly changing within the agencies during the three-month period of data collection. Since many of the questions addressed to occupants focused on evaluative ratings of the surrounding physical conditions, there was the problem of lapsed time between the collection of the evaluative or subjective measures and the collection of objective environmental data. It would have been more ideal to distribute the questionnaires and take the environmental measurements at the same time. The scope of the data collection effort relative to the availability of research personnel and the delays in obtaining necessary instruments necessitated our collecting environmental data ten weeks after the questionnaires had been completed. We know, for example, that some building occupants who had completed the questionnaire left their jobs during the interim period, and others changed the arrangement or location of their work stations. Other types of changes occurred as well. In the two-week period prior to the distribution of the questionnaires, the building's heating and ventilating system was dramatically altered and conditions were not fully stabilized when the questions about heating and ventilation were answered. Nor did the employees' responses necessarily reflect the ambient conditions that were actually measured ten weeks later. Furthermore, both data collection efforts occurred during the winter months and not also during the summer or an interseason period when ambient conditions and people's responses to them might have been different.

The second issue arising from our data collection procedures has to do with measurement precision and the amount of time devoted to completing the measurement task. As we noted above, a detailed approach to measuring the glare condition at each work station would have been difficult and expensive to perform if a high degree of precision had been required. Several other measures, such as temperature and humidity, did not vary significantly within agencies or within the building as a whole, and so our measures were taken at the agency rather than work-station level. This approach enabled us to save time, but at the cost of precision in determining ambient conditions at the individual work stations. Similarly, the total amount of time devoted to data collection did not enable us to take more than two readings of specific environmental conditions at each work station. Under ideal conditions, several measures should have been taken over a period of time to reflect variations in conditions, measures which could then be averaged to produce a single composite indicator of a specific environmental condition. Further discussion of these limitations in our data collection efforts will be included in the final chapter.

*Documentation, Data Analysis, and Dissemination*

Prior to the completion of interviews with Ann Arbor residents, work was initiated on the preparation of codebooks to be used in transferring questionnaire responses into machine-readable, quantitative form suitable for general consumption. The codebooks were basic referral documents used by the researchers in planning subsequent data analysis.

The codebooks for the three questionnaires were completed in mid-January; the coding process began shortly thereafter and lasted about two weeks. By mid-February, the initial findings were available for dissemination.

Preliminary findings from the interview phase of the study were presented in tabular and graphic form to GSA officials, their architects, and the employees in the Ann Arbor Federal Building. We recognized that presenting data that summarized all responses given by each of the three user groups would be unwieldly as a first stage in the feedback process; accordingly, we reported selected findings of a general nature along with detailed findings covering specific evaluative issues. Summary tables covered such items as the proportion of each group who (a) viewed the building favorably, (b) were satisfied with its location, and (c) felt its interior was attractive. Average ratings of specific work station characteristics were compiled for personnel in each agency and presented in graphic form. Comparisons were shown between responses in this study to specific work station characteristics and responses to identical items included in the national survey of office workers prepared by Louis Harris and Associates (1978). A summary of the documents used to convey our preliminary findings is shown in the figures in Appendix A.

Presentations of our preliminary findings were also made to several groups in the University community. We felt that, in addition to the key groups who were central to the development and use of the building, the findings should be made available to the community-at-large. Indeed, considerable interest within the community was generated as a result of local newspaper reports on the evaluation which appeared at that time.

Following these initial dissemination efforts, we arranged a series of discussions with GSA officials in Washington and with members of the architectural firm that designed the building. These groups, along with agency personnel from the Ann Arbor Federal Building, were invited to raise questions that might be answered with our data and to maintain contact with members of the research team during the remainder of the evaluative process. (We found it surprising that no one accepted the invitation.)

Environmental data were being collected and recorded at the building during the same period when preliminary findings from the user surveys were being disseminated. These environmental data were subsequently coded and merged with the data covering employee attitudes and behaviors. At this point, we were able to examine the employees' subjective responses to their environment in relation to specific environmental attributes.

The merged data set covers 220 federal employees and their work stations. Nineteen other employees responded to the questionnaire, but, because they did not have a specific work station assigned to them, their responses were excluded from the merged data set. Similarly, environmental data were obtained for 265 work stations, but only 220 cases are represented in the merged data set. The remaining 45 work stations for which data are available either did not have a federal employee assigned to them or the employee who worked there did not respond to the questionnaire. The schematic relationship between the three data sets is shown in Figure 2.2.

Beginning in mid-March of 1980, a concerted effort was begun to analyze the combined data sets and to test the hypotheses. In addition to preparing descriptive statistics covering individual responses and objective environmental conditions, we constructed a number of indexes as a means of reducing the available data. A series of both bivariate and multivariate analyses were subsequently performed. Portions of these analyses are presented in later chapters of this volume.

**Conceptual Model**

During the same period of time when we were trying to identify the kinds of data necessary to address the most salient evaluative issues and to test hypotheses, the research team also began work on the development of a conceptual model to demonstrate the manner in which interrelationships among data could be examined. It has been suggested in the literature that a weakness of previous environmental evaluations has been their lack of a conceptual framework for guiding analysis (Marans, 1978; Canter et al., 1980). Indeed, such frameworks have been lacking in most of the research dealing with people and their physical settings.

An underlying purpose of any environmental evaluation should be to develop a better understanding of how the physical environment or place contributes to or impedes the goals of the individuals or groups who must operate there. Specifically, the research should attempt to clarify and supplement what is presently known about relationships between both the physical environment and its specific attributes and people's behaviors and subjective responses to that environment. Within any environmental context, there clearly is a multitude of interrelationships which require examination if this basic objective is to be fulfilled. Certainly this is true in the case of the Ann Arbor Federal Building.

A conceptual model is presented here as a mechanism for understanding the interrelationships among data collected as part of this study. The model has served two additional purposes. First, it provides the reader with a "map," showing how different sets of variables covering federal employees and their actions, feelings, and environmental set-

FIGURE 2.2

Relationships between Data Sets

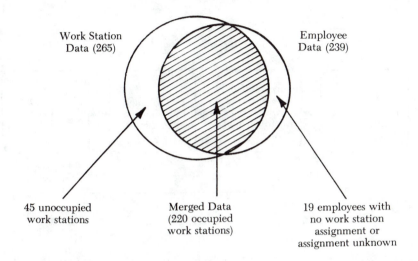

| Work Station Data (265) | | Employee Data (239) |

| 45 unoccupied work stations | Merged Data (220 occupied work stations) | 19 employees with no work station assignment or assignment unknown |

tings might be interrelated. Second, it has served as an organizational framework for guiding the data analysis.

Our conceptual model for this study was derived in part from a framework previously developed by one of the authors for use in conducting research on relationships between objective conditions, subjective experiences, and residential satisfaction (Marans and Rodgers, 1975). Basically, that model suggests that an individual's expressed satisfaction with the residential environment is dependent upon his or her evaluation or assessment of several attributes of that environment. How a person evaluates a particular attribute is in turn dependent on two factors: how that person perceives it and the standards against which he judges it. An individual's perception of a particular attribute is dependent on but distinct from the objective environmental attribute itself. The possibility of bias, inaccuracy, or simply differences in perceptions among individuals in the same environment is recognized explicitly. Finally, the characteristics of an individual are seen as affecting his perceptions and assessments of environmental attributes and the standards for comparisons that are used.

As an extension of this framework, it has been posited that satisfaction with the residential environment together with satisfaction with other domains of life can influence the quality of life as an individual experiences it. Similarly, residential satisfaction is seen as contributing both

FIGURE 2.3

Basic Conceptual Model

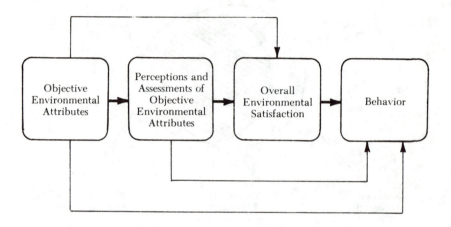

FIGURE 2.3

Basic Conceptual Model

to selected behaviors of residents and to the extent to which these behaviors occur within the residential setting.

From the perspective of the environmental designer, the core of the
model is represented by the direct and indirect links between objective
environmental attributes, people's subjective responses to these attributes, overall environmental satisfaction, and some specific behavior.[15]
These relationships, which are shown in Figure 2.3, could be applied
to the analysis of data covering a variety of environmental settings.

Of course, not every evaluation of a physical environment or place
would operate with the same set of variables. Places differ in their purposes, and the variables to be considered are usually determined after
these purposes have been identified and prioritized. Nor, for that matter, are all evaluations undertaken for the same reasons or with the
same level of funding and sophistication. Nonetheless, place evaluations conducted from the perspective of users can operate from a common analytical framework, irrespective of the type of physical environment that is being evaluated.

We noted above that the original conceptualization was developed
in conjunction with research aimed at evaluating residential environments. The model has also been used in connection with research on
recreational environments (Marans and Fly, 1981), and variations on
the model have been used to guide evaluative research in institutional
settings (Canter et al., 1980). Evaluations of each type of physical en-

vironment have operated under the assumption that any particular place is made up of component parts or environmental attributes. Furthermore, each attribute can be assessed by people who use that place, and the sum of the individual assessments contributes both to an overall evaluation of the place and to specific behaviors that take place within it. The kinds of overall evaluations and specific behaviors to be considered differ depending on the type of place being evaluated and the particular outcomes or indicators of success that are thought to be important. For example, in evaluations of residential environments, outcomes may have to do with dwelling satisfaction, neighborhood satisfaction, or the desire to move from a particular locale. In an evaluation of hospital wards, outcomes may be related to patient comfort or the ability of doctors and nurses to give care to patients.

The issue of appropriate outcomes or indicators of success in work environments has received considerable attention in recent years. At the same time, research on the quality of working life, both in office and industrial settings, has viewed the physical environment as one factor contributing to that quality. Much of this research has treated overall job satisfaction as a key outcome measure, while organizational studies of work environments have considered worker performance as an indicator of success.

In evaluations of work environments, it seems reasonable to consider both job satisfaction and job performance as appropriate outcome measures. No doubt other criteria could also be identified in evaluating any particular work setting, and their selection would generally reflect a variety of factors, including the purposes of the study, the interests of the client, who the evaluators are, who the study sponsor is, and what resources are brought to bear on the work.

Figure 2.4 graphically depicts a conceptual model for evaluating work environments. In this model, three key outcomes are suggested — overall environmental satisfaction, job satisfaction, and worker performance. As noted above, overall environmental satisfaction is the common ingredient of all place evaluations; it is the outcome of greatest interest to architects and space planners and the one receiving the most attention in this work. The model suggests the manner in which conditions or attributes of the work place are linked to the satisfaction and experiences of workers.[16]

Overall environmental satisfaction for an employee is dependent upon four factors. First, the employee's position or job type may influence how he or she evaluates a work environment. A clerical worker and a manager both working in the same open-office arrangement may have very different feelings about their work environment. Second,

FIGURE 2.4

Conceptual Model for Evaluating Work Environments

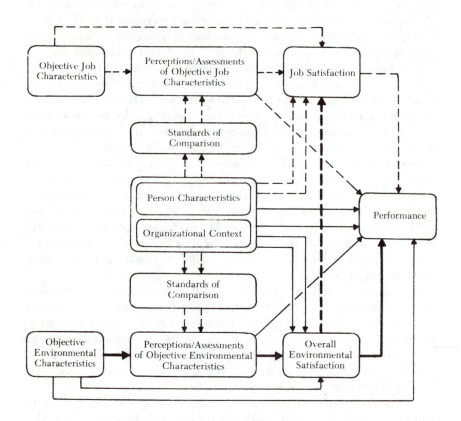

overall environmental satisfaction is dependent upon the organizational context in which employees operate. The organizational context encompasses but is not limited to the mission of the organization, the activities that take place within it, the morale of the organization, and the general nature of employee/employer relations. An employee requiring privacy may not view his or her workspace favorably if the organizational requirements also necessitate its being used for group meetings. Overall environmental satisfaction is also dependent on the individual's perceptions and assessments of several specific attributes of the physical environment. Finally, the objective attributes themselves contribute to overall environmental satisfaction. Excessive noise and stuffy air, aside from a person's perceptions of these attributes, could influence that individual's feelings about the office in which he works.

The model also shows that an individual's perception and assessment of a particular environmental attribute is dependent on two factors: the standards against which he or she judges that attribute and the objective attribute itself. The standards for comparison may include the level of a particular attribute (a) that has been previously experienced (e.g., less noise); (b) that is assigned to co-workers (e.g., closer to the boss); or (c) to which he or she aspires or expects to receive along with a promotion (e.g., more space).[17]

As we noted in our discussion of the basic model, an individual's perception or assessment of an environmental attribute is related to but distinct from the objective attribute. An employee operating in a very high-density workspace, for example, may not necessarily feel crowded or lacking in privacy. From the point of view of researchers and the environmental designer, a central purpose of evaluation research is to explore such connections between specific environmental attributes and people's perceptions of them. By understanding these relationships, the designer will ultimately be in a better position to judge the ways in which prospective users of the built environment are likely to respond.

Individual perceptions and assessments of specific environmental attributes and the attributes themselves also contribute to a worker's job performance. High noise levels and feelings about being crowded can be distracting and can affect the quality and quantity of work produced. Moreover, the characteristics of the individual and his or her organizational context are likely to have some bearing on job performance.

Another set of relationships implied by the model and suggested by the literature dealing with the quality of work life has to do with specific job characteristics as they relate to the worker's perceptions and assessments of them and to overall job satisfaction. One specific job characteristic and the responses to it centers on the quality of the physical environment. This job characteristic, represented in our model by the box labeled "Overall Environmental Satisfaction," provides a unique contribution to overall job satisfaction. Finally, job satisfaction, like job performance, is likely to be influenced by the characteristics both of the individual worker, such as age and seniority, and of the organization within which he or she operates.

While it is possible to develop appropriate measures for each element of the model within the context of any work environment evaluation, certain limitations might arise that would prevent the researchers from doing so. In this case study, no attempts have been made to measure the full range of employee job characteristics or the ways in which these characteristics were assessed by individual employees. Nor was there any effort made to measure their overall job satisfaction. In part, these limitations were imposed by individuals whose cooperation was essential to the successful completion of the research. Similar limitations were placed on the research team in our efforts to measure worker performance. Finally, the identification of specific characteristics of each organization within the building was considered to be beyond the bounds of our investigation. At best, we can differentiate between organizations by indicating the particular agency in which the individual employees worked.

## Notes

1. As noted in Figure 2.1, the intensive period of work covering the first three phases lasted about six months. During this time, five students, along with the principal investigator, were actively involved in the work. Only the two authors and a research assistant were involved in the study for the remainder of the evaluative process, data analysis, and report writing.

2. That study was prepared by Louis Harris and Associates for Steelcase, Inc. (1978).

3. The three questionnaires used in the evaluation are included in Appendix B.

4. The telephone directory used was circulated in December 1978, nearly 11 months prior to the scheduled period of interviewing. Given a large number of student households in Ann Arbor, we expected that many of the listed phone numbers would be disconnected. We also recognized that a systematic selection of lines from the pages of the directory would yield nonresidential numbers such as commercial establishments, governmental agencies, professional offices, and children's listings. We systematically selected 216 phone numbers and, of these, 174 proved to be residential telephone numbers.

5. During the 35 half-hour intervals that exist between 8:30 and noon Monday through Friday and the 25 weekly half-hour periods between 1:00 and 3:30, a sample of ten half-hour intervals was systematically selected. The results of these counts taking averages for the week are shown in Chapter 6, Figure 6.1. The number entering and leaving the building during each period was estimated at twice the number counted during the 15-minute intervals.

6. Immediately after the two observers completed their counts covering the assigned entrance, one moved to a location in the main lobby which enabled him or her to see the information desk and hear conversations that took place there while the other went to the second-floor lounge area. For 15 minutes, the first observer recorded the number of people who sought assistance from the security guard at the desk. During the same period, the second observer recorded the number of people who went into the snack bar and the number who sat in the lounge area.

7. The pre-questionnaire letter to the building occupants is shown in Appendix B.

8. It should be recognized that the number of people working in any building changes from time to time. The Ann Arbor Federal Building is no exception. During our initial meeting with GSA representatives, we were told that 292 people were working in the building. At subsequent meetings with the heads of agencies, we learned that some had vacant positions to be filled, while others had a larger staff than they had originally anticipated. In some instances, several part-time employees were filling a single position. In other words, the actual number of building occupants was constantly in flux. A final check

with agency heads just prior to distributing the questionnaire revealed that 270 people were employed in the building.

9. The total includes a few questionnaires returned during the subsequent week from employees who had been on vacation or travelling in connection with their jobs.

10. Thirty-six numbers had been disconnected, and 23 resulted in either refusals or no one at home after four call-backs.

11. The complete set of indirect environmental conditions and the form used to record indirect measures are shown in Appendix C.

12. The form used to record the direct environmental data is shown in Appendix C.

13. In a number of cases, readings were taken by two or more members of the research team so as to test the accuracy of the measurement procedures and to insure consistency.

14. See Chapter 7, Figures 7.1 through 7.6, for the locations at which recordings were made.

15. In reality, the evaluation would consider a number of behavioral outcomes which could be affected directly as well as indirectly by the objective environmental attributes and people's perceptions of them.

16. The reader will note that the figure contains continuous and broken lines differing in thickness. The heavy lines suggest relationships of importance to the environmental designer; broken lines represent relationships that could not be examined as part of this evaluation since data were not collected for several of the key elements (job characteristics, job satisfaction, standards of comparison, and so forth). Double lines denote characteristics of organizations and their individual employees.

17. The concept of a standard of comparison is a complex one and is often difficult to measure within the context of evaluation research. For a more thorough discussion of the nature of these standards, see Campbell et al. (1976).

# 3

# The Ann Arbor Federal Building

As a major educational and research center in the Great Lakes region, the city of Ann Arbor has often been considered a desirable location for federal agencies. In coming to the city, most of them have opened offices in leased commercial space in downtown Ann Arbor or in the outlying areas. In the early 1970s, Ann Arbor was designated as the site for a new federal district court, setting the stage for the consolidation of agencies of the federal government into a single location. After funds were appropriated, the General Services Administration was directed by Congress to build a structure in Ann Arbor that could accommodate a number of governmental units, including court facilities. The decision was made within GSA to design and construct a building that would not only accommodate various administrative, postal, and judicial functions, but that would also be flexible enough to accept these various activities over a period of time. A major goal was to create a facility of the size and configuration necessary to house numerous agencies whose organizational arrangements and spatial needs varied widely.

During this period, new guidelines for the design of federal buildings were being established by GSA. New federal structures were to be unlike the stereotypically large and impersonal buildings characteristic of public architecture of the early twentieth century; new buildings were to be responsive to the surrounding urban environment and located in close proximity to public transportation, and they were to be designed to reflect the federal government's growing concern for energy conservation.

FIGURE 3.1

The Setting

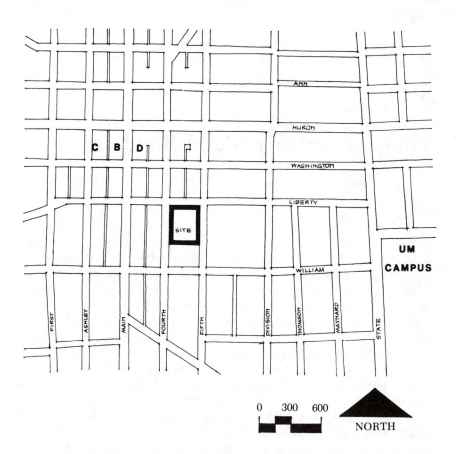

In early 1973, GSA announced that the new federal building would be built in downtown Ann Arbor on the northern half of the block bounded by Liberty and William Streets and Fourth and Fifth Avenues (see Figure 3.1). To the city's planning commission this was not the first, nor even the second, priority site, and it meant having to raze the historic Masonic Temple. But officials of the federal government maintained in their environmental impact statement that this site was the most advantageous in terms of their building objectives. The site was well situated with respect to Ann Arbor's central business district and to the University of Michigan campus; it was close to the central bus terminal and public transportation lines serving the Ann Arbor area;

and a building on this site was expected to be a catalyst for new development in downtown Ann Arbor. GSA selected the architectural firm of Tarapata, MacMahan, and Paulsen of nearby Bloomfield Hills, Michigan, for the project.

Community resistance to the building and its location was triggered by the proposal to demolish the Masonic Temple to accommodate 40 parking spaces. The Citizens' Association for Area Planning (CAAP) proposed several alternative sites in an attempt to save the temple, and they suggested that another story be added to the existing parking structure on Fourth and William Streets. Other members of the community objected to the building because the proposed site lacked adequate parking and would result in increased traffic congestion on Fourth and Fifth Avenues. Still another concern centered on the proposed building's threat to the scale and small-town character of downtown Ann Arbor. All of these concerns were expressed at public hearings during the course of the planning process. GSA officials continued to argue that the Liberty Street site was most suitable in light of their own building philosophy, and the location was eventually approved by the city planning commission and by the Ann Arbor city council. Demolition of the Masonic Temple and other buildings on the site began in 1974, and by early 1976 construction of the new building had begun. The new Ann Arbor Federal Building was occupied by its first tenants in July 1977.

**The Site**

As Figure 3.2 shows, the one and a half acre site of the Ann Arbor Federal Building is situated in the middle of the major commercial area of the city and close to the central campus area of The University of Michigan, the city's main library, and the YMCA. The building is oriented to Liberty Street, a major vehicular and pedestrian thoroughfare connecting the commercial and academic centers of town. Directly to the west of the building on Fourth Avenue is the central public transit (bus) terminus. Fourth Avenue handles two-way traffic, while Fifth Avenue to the east is the primary one-way southbound thoroughfare for traffic leaving the downtown area.

Forty assigned parking spaces for employee use were constructed to the south of the building. To the east, there are 15 short-term spaces, including one for handicapped drivers. These are intended for use by Post Office patrons.

Post Office employees have access to a leased parking lot to the west of the building, while other building users (federal employees and the

FIGURE 3.2

Surrounding Land Use

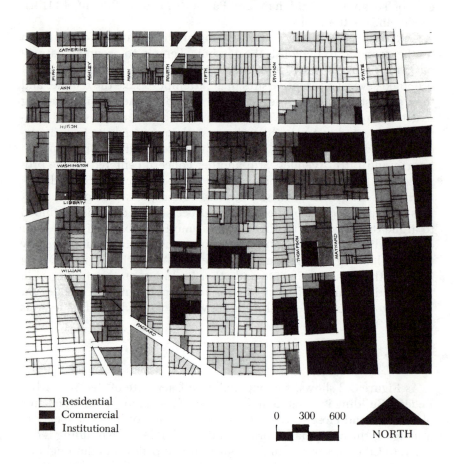

☐ Residential
■ Commercial
■ Institutional

0    300   600

NORTH

public) have access to a nearby municipal parking structure, a large municipal lot, and street parking along Liberty and Main Streets and on Fourth Avenue, north of Liberty. Both the location of these parking areas and the traffic pattern around the building are shown in Figure 3.3.

## Design Concept

A major theme of the design concept for the building is the orientation toward Liberty Street. The building faces a large open plaza extending the length of the Liberty Street facade and provides a break in

FIGURE 3.3

Parking and Transportation

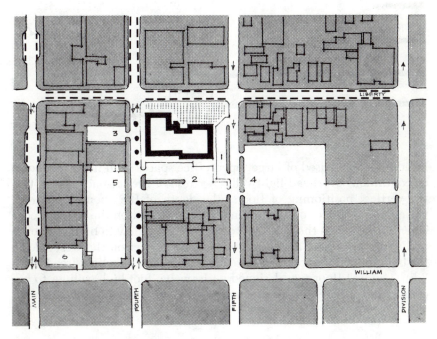

| | | |
|---|---|---|
| ••• Public transit terminus | 3 | Postal workers parking lot |
| − − − Metered parking | 4 | Municipal parking lot |
| 1 Short-term parking | 5 | Municipal parking structure |
| 2 Employee parking lot | 6 | Municipal parking lot |

the typical ten-foot setback of adjacent buildings. The architects further decreased the mass of the Liberty Street facade by stepping each successive story of the four-floor structure back to the south and opening the north face with a continuous wall of glass (see Figure 3.4). The east and west facades are windowless, except for a small window on the first floor, and are thus closed to exterior views. The south side is open only on the first floor to a Post Office service deck and a pedestrian entry ramp; it also has small windows on the fourth floor. The entire exterior is composed of light terra-cotta tile and patent glazing.

Except for the Post Office on the ground floor, the interior of the

FIGURE 3.4

Design Concept

building is comprised of large, open-office spaces with north windows or continuous overhead lightwells. Because of the set-back concept in the vertical positioning of floors, the depth of the open-office areas varies from 150 feet on the ground floor to 40 feet on the fourth floor. The fourth floor is the only one to have a direct view to both the north and south. All floors are connected to one another on the west side of the building with an open lightwell located below a skylight (see Figure 3.5). The first floor was designed with twelve-foot high ceilings to accommodate a federal district courtroom facility at some future date.

FIGURE 3.5

Conceptual Flow of Space between Floors

*Ann Arbor Federal Building — north side. A major element of the design concept is the orientation of the building toward Liberty Street and the plaza on the north.*

The building is considered by many to be innovative in its design, and it has received considerable publicity in the architectural press. It has also been the recipient of numerous honors, including design awards from the Detroit Chapter of the American Institute of Architects and the Michigan Society of Architects.

## The Agencies

During the period of the evaluation, the Ann Arbor Federal Building housed 14 federal agencies employing a total of 270 to 292 federal workers.[1] Below is a brief description of each of these agencies (as of November 1979):

*U.S. Post Office.* Located in the eastern half of the building on the ground floor and with entrances along Liberty Street and from the main lobby of the building, this agency employed 70 postal workers, including counter clerks, postal carriers, and supervisory personnel. It had previously been housed in an old Post Office building on Main Street at the northern edge of downtown Ann Arbor.

*Internal Revenue Service (IRS).* This agency employed 66 tax agents and clerical workers. It is situated on the northwest part of the ground floor, with entrances off the main lobby. It has one closed skylight on the north wall, an open lightwell in the center of its space, and a small west-facing window. All office space is arranged with moveable, five-foot high partitions. The agency had previously leased space in a commercial shopping center about two miles southeast of downtown Ann Arbor.

*Army, Air Force, Marine, and Navy Recruiters (Military Recruiters).* These agencies maintain four separate offices on the southwest corner of the ground floor, with entrances to the south lobby. Fifteen enlisted military personnel and two civilian receptionists occupied these four, glass-enclosed offices. The offices of the Army, Air Force, and Navy recruiters each have two small, south-facing windows. Prior to their move to the federal building, the military recruiters leased ground-floor commercial space in separate locations in downtown Ann Arbor. The offices of the military recruiters, the postal employees, and IRS are shown in Figure 3.6.

*Heritage Conservation and Recreation Service (HCRS).* This regional office of HCRS is located on the western half of the second floor, with entrances to a lobby that also serves as an employees' lounge area. Approximately 41 staff members were employed by the agency during the study. The entire north wall is a full-height window and open lightwell; the office space is bisected by a continuous open lightwell at the

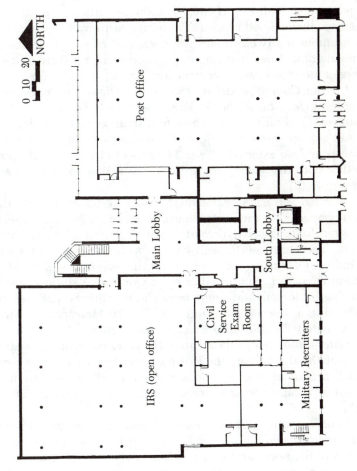

FIGURE 3.6

Ground Level Floor Plan

entry. All open offices have five-foot high moveable partitions. Prior to their move, HCRS was housed in leased space in an industrial park four miles south of its present location.

*Social Security Administration.* Occupying the northeast portion of the second floor, this agency is connected to the second-floor lobby by elevators and a stairway. The agency employed 33 case workers and clerical personnel who, for the most part, were housed in an office-pool environment. Three private, windowless offices are located to the south of the main office area. The entire north wall is glass and a closed, continuous lightwell bisects the open and closed office areas. Before moving downtown, the agency was housed in a private office building three miles northeast of central Ann Arbor.

*District Court-Probation Office, the Defense Investigative Service, the Department of Labor — Wage and Hourly Division, and the Army Surgeon General's Office.* These four small agencies are located on the south side of the second floor. Each has two employees with private offices and no external views. The offices of these small agencies, together with the Social Security Administration and HCRS, are shown in Figure 3.7.[2]

*Weather Service.* Located on the northwest portion of the third floor and connected to the central elevator lobby, this agency operates on a 24-hour basis. Twelve forecasters are on duty during any given 8-hour shift. The entire north wall has an open lightwell and a full-height window. One private office exists for the director, while most of the remaining space is comprised of open work areas situated around forecasting equipment and computers. The Weather Service was previously housed in a weather data station at Detroit Metropolitan Airport, approximately 25 miles to the east of Ann Arbor.

*Federal Bureau of Investigation (FBI).* Located on the southeast portion of the third floor, the FBI has seven agents and one secretary in small private offices with no windows. The locations of both the FBI and the Weather Service are shown in the third-floor plan of Figure 3.8.

*Department of Agriculture-Soil Conservation Service.* Housing three staff members on the west end of the fourth floor, this agency has open views to the north and south and open-office furniture arrangements.

*Department of Defense-Army Recruiting Area Commander.* Three military personnel are located in a single private office on the fourth floor; there is a window on the south wall.

*Defense Logistics Agency.* This agency houses three civilian staff workers in a private fourth-floor office with no external views. The latter three agencies are shown in the fourth-floor plan in Figure 3.8.

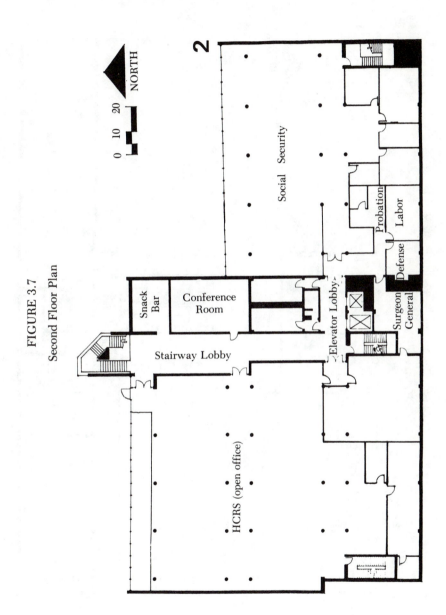

FIGURE 3.7
Second Floor Plan

FIGURE 3.8

Third and Fourth Floor Plans

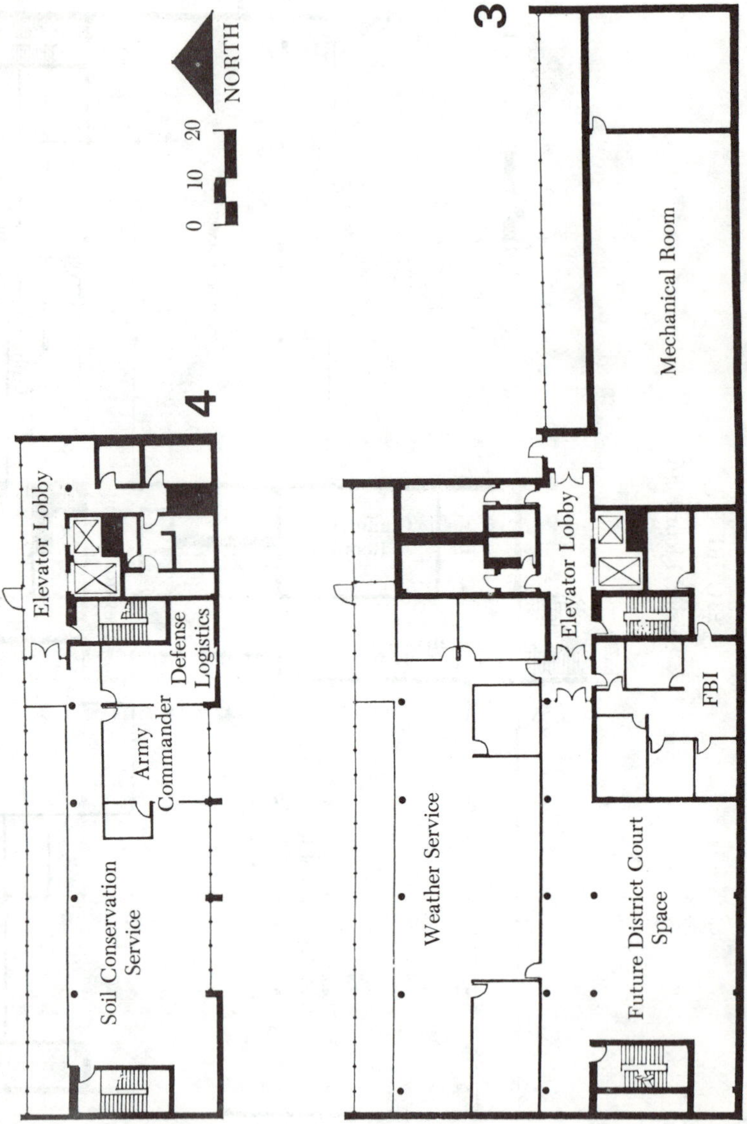

*Federal Building — south side. A parking area and a wall of terra-cotta
with few windows characterize the south side of the building.*

In addition to the offices of these agencies, the building contains a small office in the basement used by the security guards, a conference room on the second floor, a snack bar next to the conference room, and a large room adjacent to the military recruiters; this room was used for testing by the U.S. Office of Personnel Management (formerly the U.S. Civil Service Commission) and was occupied for approximately three hours a week.

**The Building Users**

The agencies housed in the Ann Arbor Federal Building serve the public in a variety of ways and to varying degrees. Some are visited daily by hundreds of customers or clients, while others are barely known to the public. People go to the building to purchase postage stamps or pick up packages, to apply for social security, or have their tax returns audited. Others come to simply seek out information from agencies such as the IRS, the HCRS, or the Soil Conservation Service. To some extent, the varied functions and services offered reflect the diversity of clients served. The occupants of the building also vary, in terms of their professional training, their job classifications, and the length of time they have worked for the federal government.

As noted earlier, 270 employees worked in the agencies at the time of our evaluation. Two-thirds of them were male and they performed a wide range of professional or technical jobs. The average employee had been with his or her respective agency for seven and a half years. Most worked in the building five days a week, although a number worked outside of the building part of the time. Approximately two-thirds of the employees' typical work day was spent at their desks or work stations. Employees had an average of seven daily contacts at their desks with either co-workers or members of the public, and they spent an average of nearly an hour and 15 minutes on the telephone each day.

Most public users of the building were residents of Ann Arbor, and approximately one in five were students at The University of Michigan. About half of the residents contacted were employed downtown or at the University campus and were frequent users of the downtown facilities, including the Federal Building.

The day-to-day management of the building required little supervision. General problems and maintenance were handled by two federal security guards and privately contracted maintenance personnel. The overall management was under the auspices of the General Services Administration, whose building manager visited the site two or three times each month or in connection with specific problems.

## Who works in the building?

Nearly 90 percent of the 270 employees in the Ann Arbor Federal Building responded to questions designed to characterize their jobs and tap their thoughts and activities relative to the building. Table 3.1 deals with a number of the employee characteristics. About 60 percent of those who responded were in managerial or professional-technical jobs, 15 percent indicated they had clerical or secretarial positions, 19 percent were postal carriers, and the remaining 5 percent were military recruiters. Thus, except for the Postal Service, the federal agencies were staffed by large numbers of professional and technical personnel, and about two-thirds of these jobs were held by men. In HCRS, the Social Security Administration, and the small agencies, one in five employees indicated they held clerical-secretarial positions; virtually all of these positions were occupied by women (87 percent). Among all the employees in the building, one-third (35 percent) were women.

With the exception of the military recuiters, most employees had worked in the Ann Arbor Federal Building for at least one year prior to the administration of the questionnaire. On average, people had worked in the two-year old building for a year and a half; this would suggest that most of the employees had had ample time to experience their work environment and the operation of the building during different seasons. Indeed, 44 percent of the employees had worked in the building since it opened in the fall of 1977.

## What kinds of work do they do?

The range of agencies and the types of jobs within each suggest that the specific tasks of the federal employees in the building varied considerably from agency to agency. As part of our environmental data collection, we attempted to identify job tasks at each work station in that building. For example, interviews with clients and customers took place at one-quarter of all work stations, while only three work stations (one percent) had a computer terminal. Most work stations or desks (82 percent) were used by people engaged in writing, filing, or other types of clerical tasks.

We also asked people about their work schedules, the amount of time they spent at their desks or work stations, their meeting schedules, and their telephone activity. Table 3.2 shows that most of the employees (81 percent) spent at least five days a week in the building. People working less than five days per week at the building were most likely to be employed in IRS, HCRS, and in the small agencies.

On average, federal employees spent approximately two-thirds of

TABLE 3.1

Employee Characteristics, by Agency
(Percentage Distribution)

| Employee Characteristics | All | Agency | | | | | | |
|---|---|---|---|---|---|---|---|---|
| | | Post Office | IRS | Military Recruiters | HCRS | Social Security | Weather Service | Small Agencies[a] |
| Job Classification | | | | | | | | |
| Manager-supervisor | 11 | 6 | 6 | 8 | 15 | 9 | 25 | 22 |
| Professional-technical | 50 | – | 80 | – | 64 | 73 | 69 | 52 |
| Clerical-secretarial | 15 | 8 | 14 | 8 | 21 | 18 | 6 | 22 |
| Postal carrier | 19 | 86 | – | – | – | – | – | – |
| Military recruiter | 5 | – | – | 84 | – | – | – | 4 |
| Total | 100 | 100 | 100 | 100 | 100 | 100 | 100 | 100 |
| Number of respondents | 235 | 51 | 49 | 12 | 47 | 33 | 16 | 27 |
| Sex | | | | | | | | |
| Female | 35 | 20 | 35 | 8 | 41 | 81 | 13 | 22 |
| Male | 65 | 80 | 65 | 92 | 59 | 19 | 87 | 78 |
| Total | 100 | 100 | 100 | 100 | 100 | 100 | 100 | 100 |
| Number of respondents | 234 | 54 | 48 | 12 | 46 | 32 | 15 | 27 |

| | | | | | | | | |
|---|---|---|---|---|---|---|---|---|
| **Length of Time with Agency** | | | | | | | | |
| Less than one year | 12 | 7 | 14 | 16 | 19 | 9 | 13 | 8 |
| 1–2 years | 11 | – | 10 | 23 | 17 | 9 | – | 22 |
| 2–5 years | 18 | 15 | 21 | 15 | 15 | 31 | 7 | 15 |
| 5–10 years | 30 | 39 | 37 | 15 | 21 | 27 | 40 | 22 |
| More than 10 years | 29 | 39 | 18 | 31 | 28 | 24 | 40 | 33 |
| Total | 100 | 100 | 100 | 100 | 100 | 100 | 100 | 100 |
| Mean time with agency (years) | 7.5 | 9.3 | 6.4 | 6.7 | 6.7 | 6.9 | 9.3 | 7.5 |
| **Length of Building Occupancy** | | | | | | | | |
| 6 months or less | 19 | 15 | 27 | 46 | 11 | 18 | 25 | 14 |
| More than 6 months; less than one year | 10 | 7 | 6 | 23 | 11 | 9 | – | 23 |
| 1–2 years | 27 | 26 | 22 | 31 | 23 | 27 | 62 | 19 |
| More than 2 years | 44 | 52 | 45 | – | 55 | 46 | 13 | 42 |
| Total | 100 | 100 | 100 | 100 | 100 | 100 | 100 | 100 |
| Mean length of occupancy (years) | 1.5 | 1.6 | 1.4 | 0.8 | 1.6 | 1.5 | 1.3 | 1.4 |
| Number of respondents | 238 | 54 | 49 | 13 | 47 | 33 | 16 | 26 |

a Includes Department of Defense–Army Recruiting Area Commander, Defense Logistics Agency, Defense Investigative Service and Army Surgeon General; Soil Conservation Service; District Court-Probation Department; Department of Labor, Wage and Hourly Division; the Federal Bureau of Investigation, and the Security Guard.

TABLE 3.2

Work Schedule and Time at Desk/Work Station, by Agency
(Percentage Distribution)

| Work Schedule | | Agency | | | | | | |
|---|---|---|---|---|---|---|---|---|
| | All | Post Office | IRS | Military Recruiters | HCRS | Social Security | Weather Service | Small Agencies |
| Days per Week at Building | | | | | | | | |
| 2 days or less | 7 | 2 | 18 | – | 4 | – | 6 | 19 |
| 3–4 days | 12 | – | 27 | – | 13 | 9 | – | 22 |
| 5 days | 74 | 98 | 55 | 42 | 81 | 76 | 94 | 48 |
| More often | 7 | – | – | 58 | 2 | 15 | – | 11 |
| Total | 100 | 100 | 100 | 100 | 100 | 100 | 100 | 100 |
| Number of respondents | 238 | 54 | 49 | 12 | 47 | 33 | 16 | 27 |
| Percentage of Time at Desk/Work Station[a] | | | | | | | | |
| 100 percent | 3 | 4 | 4 | – | 2 | 3 | 6 | 4 |
| 76–99 percent | 49 | 25 | 49 | 41 | 57 | 67 | 69 | 52 |
| 51–75 percent | 19 | 9 | 16 | 42 | 32 | 18 | 19 | 11 |
| 50 percent or less | 29 | 62 | 31 | 17 | 9 | 12 | 6 | 33 |
| Total | 100 | 100 | 100 | 100 | 100 | 100 | 100 | 100 |
| Mean percentage of time at desk/work station | 63 | 46 | 63 | 66 | 72 | 73 | 77 | 62 |
| Number of respondents | 237 | 53 | 49 | 12 | 47 | 33 | 16 | 27 |

[a] The question was: "On an average work day, about how much of your time is spent at your desk or work station?"

their day at their work stations. Not surprisingly, the postal workers were the most likely to spend time away from their work stations; employees in the Social Security Administration and the Weather Service were most likely to work at their desk.[3]

There was considerable variability in the extent to which employees interacted with others at their work stations (Table 3.3). On average, the federal employees met twice a day with outsiders and contacted their co-workers more than four times daily. Most likely to meet clients or customers were the military recruiters on the first floor and the staff of the Social Security Administration on the second floor. However, the most frequent contacts with the outsiders were made by the clerks in the Post Office who operated the customer service counter. HCRS personnel and employees in the Weather Service were least likely to meet with outsiders.

Interaction with co-workers was most likely to occur among the military recuiters and the Social Security employees, but there was also a high level of co-worker interaction among Weather Service personnel. Half of the employees in the latter agency indicated they met with fellow workers at least five times during a typical day. This intensive interaction is not surprising since the mission of the Weather Service requires a constant exchange of weather forecasting information. Least likely to meet with co-workers were the postal workers, who for the most part carried out their tasks independent of one another. A composite measure of meetings with both co-workers and members of the public was created and is shown in the last part of Table 3.3. It can be seen from these data that the military recruiters (10.1 contacts) and Social Security Administration personnel (9.7 contacts) were most actively involved in interchanges at their respective work stations. In subsequent chapters, we will see if and how this activity influenced a number of attitudinal and behavioral responses of employees.

Telephone use in the building also varied among agency personnel (see Table 3.4). On average, federal workers had 6 telephone conversations each day. The number of calls varied from less than 2 per day among postal workers to 14 for military recruiters. Most telephone conversations were reported by people occupying private offices (over 10 call per day), while employees in open offices made just over 6 calls during the average day.

*Who are the customers and community residents that use the Ann Arbor Federal Building?*

Both our systematic and impressionistic observations revealed a heterogeneous group of outside users. Men and women of all ages and

TABLE 3.3

Frequency of Meeting with Others at Desk/Work Station, by Agency (Percentage Distribution)

| Frequency of Meetings | All | Agency | | | | | | |
| --- | --- | --- | --- | --- | --- | --- | --- | --- |
| | | Post Office | IRS | Military Recruiters | HCRS | Social Security | Weather Service | Small Agencies |
| With Outsiders[a] | | | | | | | | |
| None | 51 | 73 | 38 | 17 | 65 | 39 | 69 | 30 |
| 1–2 times | 29 | 17 | 42 | 8 | 33 | 12 | 25 | 56 |
| 3–4 times | 8 | 2 | 10 | 58 | – | 15 | 6 | – |
| 5–10 times | 5 | 2 | 2 | – | 2 | 21 | – | 7 |
| More than 10 times | 7 | 6 | 8 | 17 | – | 13 | – | 7 |
| Total | 100 | 100 | 100 | 100 | 100 | 100 | 100 | 100 |
| Mean number of times | 1.9 | 1.2 | 2.1 | 4.2 | 0.6 | 3.8 | 0.6 | 2.2 |
| With Co-workers[b] | | | | | | | | |
| None | 6 | 13 | – | 8 | 4 | – | – | 11 |
| 1–2 times | 31 | 49 | 37 | 17 | 19 | 15 | 25 | 37 |
| 3–4 times | 31 | 17 | 35 | 33 | 43 | 43 | 25 | 22 |
| 5–10 times | 21 | 11 | 15 | 25 | 26 | 30 | 37 | 19 |
| More than 10 times | 11 | 10 | 13 | 17 | 8 | 12 | 13 | 11 |
| Total | 100 | 100 | 100 | 100 | 100 | 100 | 100 | 100 |
| Mean number of times | 4.4 | 3.2 | 4.5 | 5.3 | 4.8 | 5.4 | 5.7 | 4.1 |
| Number of respondents | 237 | 53 | 49 | 12 | 47 | 33 | 16 | 26 |
| Mean number of work station visits per day | 6.6 | 4.2 | 7.0 | 10.1 | 5.6 | 9.7 | 6.4 | 6.7 |
| Number of respondents | 231 | 51 | 47 | 12 | 45 | 33 | 16 | 27 |

[a] The question was: "On an average working day, how often does someone from outside the building come to see you on business?"
[b] The question was: "On an average day, how many times do you meet with fellow workers at your desk/work station to discuss or perform work?"

TABLE 3.4

Telephone Use, by Agency (Percentage Distribution)

| Telephone Use | All | Agency | | | | | | |
| --- | --- | --- | --- | --- | --- | --- | --- | --- |
| | | Post Office | IRS | Military Recruiters | HCRS | Social Security | Weather Service | Small Agencies |
| Frequency of Telephone Conversations[a] | | | | | | | | |
| None | 15 | 56 | — | — | — | 9 | — | 11 |
| 1–2 times per day | 18 | 32 | 17 | — | 28 | 6 | 12 | — |
| 3–4 times per day | 17 | 2 | 31 | — | 28 | 9 | 38 | 8 |
| 5–10 times per day | 28 | 4 | 38 | — | 38 | 52 | 25 | 23 |
| More than 10 times daily | 22 | 6 | 14 | 100 | 6 | 24 | 25 | 58 |
| Total | 100 | 100 | 100 | 100 | 100 | 100 | 100 | 100 |
| Mean number of telephone conversations (per day) | 6.0 | 1.7 | 6.3 | 14.0 | 5.1 | 7.7 | 6.9 | 10.1 |
| Percentage of Time on Telephone[b] | | | | | | | | |
| More than 75 percent | 3 | — | 2 | — | 2 | 6 | — | 11 |
| 51–75 percent | 5 | — | 2 | 42 | — | 6 | 6 | 15 |
| 26–50 percent | 13 | — | 17 | 42 | 11 | 18 | 19 | 11 |
| 11–25 percent | 27 | 8 | 35 | 16 | 40 | 37 | 25 | 22 |
| 1–10 percent | 43 | 59 | 44 | — | 47 | 30 | 50 | 33 |
| None | 9 | 33 | — | — | — | 3 | — | 8 |
| Total | 100 | 100 | 100 | 100 | 100 | 100 | 100 | 100 |
| Mean percentage of time on telephone | 18 | 4 | 18 | 45 | 15 | 24 | 18 | 29 |
| Mean time per day on telephone (minutes)[c] | 75 | 18 | 77 | 153 | 68 | 110 | 80 | 117 |
| Mean length of telephone conversation (minutes) | 11.6 | 5.3 | 13.7 | 11.1 | 16.1 | 12.5 | 14.6 | 9.8 |
| Number of respondents | 235 | 52 | 48 | 12 | 47 | 33 | 16 | 27 |

[a] The question was: "On an average working day, about how many phone conversations do you have?"
[b] The question was: "On an average working day, about how much of the time is spent talking on the telephone?"
[c] Estimates are based on the amount of time during an average work day the respondent was in the building. Two questions were used to make this estimate: one dealing with the amount of time spent at the desk and the other covering the number of times an employee leaves the building in connection with work.

races were seen entering the building and visiting the various agencies. Among the people contacted by telephone and in connection with the on-site interviews, about one in five was a University of Michigan student. Everyone contacted by telephone lived in Ann Arbor, while one-quarter of the people interviewed at the building said they lived outside the city. Telephone respondents had lived in the city for more than 15 years and those contacted on-site had been in Ann Arbor for an average of 17 years. Somewhat more than half from each group worked in the city and a quarter indicated they were employed in either downtown Ann Arbor or at the University. Most visitors said they were frequent users of other downtown facilities. On average, those interviewed by telephone had been to downtown Ann Arbor 10 times during the past month, and people contacted at the building averaged 13 visits during the month to the central area.

### Who manages and maintains the Federal Building?

The building is managed and operated by the General Services Administration's regional office located in Battle Creek, Michigan. A GSA building manager from that office is responsible for the operations of the Ann Arbor Federal Building and visits the facility two to three times per month.

Two to three custodial personnel, who have a contract for this service with GSA, are responsible for the daily cleaning of the building during the late afternoon and early evening hours. Security is provided by two federal guards, who have a single desk in a small office in the basement of the building. Most of their time, however, is spent at the main lobby information desk or patrolling the interior public areas and the parking area.

### Notes

1. At the time initial contacts were made with GSA about conducting the evaluation, we were informed that 292 employees worked in the building; but this number was constantly changing due to departures, new hires, temporary part-time personnel, and so forth. At the time the questionnaires were distributed to employees (in late November 1979), 270 people were employed in the building.

2. These four small agencies, together with all the remaining agencies excluding the Weather Service, employed a total of 26 employees. In subsequent chapters, data covering these agencies and their personnel are combined.

3. An examination of the time spent at the desk for people occupying various types of work stations indicated little difference. People in an open office spent 71 percent of their time at the desk, while people in private conventional offices devoted 61 percent of their time to desk work.

# 4

# Design Objectives and Evaluative Issues

The initial phase of our approach to building evaluations included a series of meetings with individuals involved in the inception, design, and management of the Ann Arbor Federal Building. As part of these meetings, efforts were made to learn why the building had been initiated and what specific objectives GSA representatives and their architects intended to fulfill through their site selection and building design. By understanding these objectives, we expected to be better able to focus our evaluation on the most salient issues and subsequently draw conclusions about the degree to which the building has been successful. This chapter outlines the objectives of the building, as gleaned from the initial meetings and background documents. The central issues addressed in this evaluation are also discussed here. These issues have formed the basis for the analysis discussed in subsequent chapters.

## Design Objectives

Information about the purposes and objectives of a particular built environment can usually be found in the written program document that is developed from early discussions between the client and the designer/planner. Often, however, the stated objectives are limited in coverage, are not well articulated, or are missing altogether. In such instances, the purposes and objectives must be inferred from the recollections of the clients and designers and from historical documents.

Since a written program for the Ann Arbor Federal Building project was unavailable to the research team, a listing of several objectives was developed in an attempt to summarize the major objectives of the building. These statements were subsequently reviewed by GSA officials and the architects and modified:

1. The building should be an integral part of downtown Ann Arbor. It should be in visual harmony with the character of the downtown setting and a catalyst for new downtown development.

2. Interaction between building occupants and patrons and the downtown community should be fostered. It should be functionally a part of downtown Ann Arbor and should be used extensively by community residents. It should be a stopping point for pedestrians who travel along Liberty Street between downtown Ann Arbor and the University of Michigan campus.

3. The building should exemplify good architectural design without being a dominating or imposing structure.

4. The work spaces within the building should allow for flexibility and change, both within agencies and throughout the building as a whole. Flexibility should be accomplished without hindering the performance of workers. The structure should be designed to eventually house a federal district court facility.

5. The building should be designed so as to create a sense of community among the people who work there.

6. Employees should take pride and find satisfaction in their work environment.

7. Opportunities should be provided for employees to store personal belongings and personalize their work spaces according to individual tastes and interests. Work areas should be functionally efficient and conducive to agency work requirements.

8. The building should be designed as an energy-efficient structure. It should be oriented so as to take advantage of the natural lighting on the north and to minimize heat gain on the east and west.

9. Materials should be selected so as to inhibit vandalism and reduce maintenance costs.

## Evaluative Issues

Working from the statement of objectives and other information gleaned from the meetings and visits to the building, members of the research team decided to conduct the evaluation around four key issues: (1) relationships between the building and the community; (2) transportation and parking; (3) people's assessments of the building and their work environment; and (4) relationships between the work environment and worker performance.

### Building and Community Interaction

Following their decision to locate the building in downtown Ann Arbor, officials at GSA and their architects agreed that its design should be harmonious with the scale and general character of the surrounding area. It was their hope that community residents would use the building frequently because of its central location and that federal employees would become active users of downtown facilities. To test these suppositions, we needed to include questionnaire items to tap people's views on the manner in which the building was functionally and aesthetically integrated into downtown Ann Arbor.

## Transportation and Parking

A second set of issues gleaned from the early discussions touched on the problems faced by federal employees and building patrons in travelling to the building and, for those who drove, the difficulties they had in finding adequate and reasonably priced parking. The parking problem was particularly acute for federal employees, many of whose agencies had previously been located in the outlying sections of Ann Arbor where parking was readily available and free. As part of the evaluation, we systematically examined how the federal employees traveled to work and whether their mode of transportation had changed since their move to the new building. We also examined the means of travel used by the public and asked all drivers about the parking situation.

## Environmental Assessments

A central purpose of this evaluation was to consider people's assessments of the built environment at four levels. First, consideration had to be given to how people in the community actually felt about the building's location and how employees rated that location relative to where they had worked in the past. Second, the building as an architectural entity was viewed as an important object for assessment both by Ann Arbor residents and federal employees. While many aspects of the building were considered to be worthy of examination, aesthetic quality was a primary interest of both the architects and members of the research team. Third, consideration was to be given to the environment of each agency occupying the building. Specifically, the evaluation was to examine how employees felt about the layout and the appearance of their agencies. Finally, it was decided that a major focus of the study would be the individual employee's workspace or work station. As we have noted, the overall building design was based on an open-office landscape concept that resulted in new and different work arrangements for most workers. The extent to which employees viewed these new arrangements positively or negatively would be examined along with their assessments of the work station relative to what they had in the past.

*Worker Performance*

One recurring topic of conversation during our discussions with agency personnel was the "performance" of workers in the new building. This concept was never clearly defined, however, and, despite the fact that other researchers have had difficulty in measuring job performance in past studies of office settings, the research team agreed that efforts should be made to examine it vis-à-vis the physical setting. Ideally, the number of "units" processed within each agency would have been measured and compared with past records indicating units processed in the prior setting. Unfortunately, it was impossible to find a common unit for measuring performance among all agencies.

We had also considered asking people about their health and the extent to which it affected their time away from the job. However, such questions are of a highly sensitive nature and threatening to some employees. Data obtained by asking agency heads about staff absenteeism would be too general and would not enable us to examine individual worker performance relative to specific attributes of that individual's work environment. Therefore, it was decided that performance would be measured indirectly by considering selected perceptions and evaluations of the workers. Three types of questions were to be asked. First, we wanted to consider how efficient or productive employees believed they were in the new setting and whether their performance had improved or declined as a result of their move. Second, we wanted to ask each employee about the performance and efficiency of other people in his or her own agency. And finally, we wanted to ask about ambient environmental conditions and the extent to which they were bothersome or disruptive to job performance. The degree to which performance was adversely affected would be inferred by the magnitude of complaints. At the very least, we would have some indication of the ways people responded to selected attributes of their agencies.

Working around these issues, we developed specific questions directed toward the employees and outside users of the building. The same issues also formed the basis for determining the specific hypotheses to be examined as part of the data analyses. The questions and hypotheses are dealt with in subsequent chapters.

# 5

# The Building and Its Surroundings

## Overview

As we noted earlier, citizen concern was expressed at the time plans for the building were announced as to how successfully the new structure could be integrated into the scale and fabric of downtown Ann Arbor. Some concerns were raised by groups whose primary interest was in preserving the Masonic Temple on Fourth Avenue. Other groups had legitimate interests in maintaining the small-town character of the central business district and believed the new building would pose a threat to that character. Indeed, the image conjured by the proposed building was that of an all-imposing structure characteristic of federal office buildings built during the 1920s and '30s.

Proponents of the building, on the other hand, argued that the new structure would enhance the character of downtown Ann Arbor by providing an important visual element on Liberty Street—the main artery connecting the shopping areas of Main and State Streets. At the same time, its presence would be a catalyst for new downtown development and generate additional business activity by bringing more people into the area.

Concerns were also expressed about transportation to and from the building and about parking. For example, it had been suggested that the Federal Building was a major contributor to the traffic congestion in downtown Ann Arbor. As part of their central area planning activities, the staff of the city's planning department was interested in knowing about the number of people who drove to the Federal Building and where they parked. Indeed, the parking situation, we found, was a major source of the federal employees' dissatisfaction with the building. It was frustrating, too, for visitors who drove to the Post Office and other agencies and who had little time to waste in attempting to find a parking place.

From the point of view of both community residents and federal employees questioned in this study, the building has been successfully integrated into Ann Arbor's downtown. Most residents knew where it was located and had been there at one time or another. For the most part, they considered the building an attractive addition to the downtown area. Building occupants were also inclined to give high marks to the location, although the new location was not viewed as favorably by some as their previous agency location. Employees' feelings about the location were influenced largely by the extent to which they used downtown facilities. People who increased their use of nearby shops, restaurants, and banks as a result of the move were most likely to be satisfied with the building's location, and most federal employees did say they used downtown facilities more often since moving into the building.

Among the people who visited the Federal Building, most arrived by automobile. Nonetheless, significant numbers of both agency patrons and federal employees walked or came by bus. As a result of the move to the new building, about one worker in eight changed his travel mode from driving to car pooling, while an equal proportion gave up the automobile in favor of public transportation or walking.

Most employees felt that the building is conveniently located with respect to travel. But a substantial number (one in five) said that getting there is inconvenient. Feelings about inconvenience were likely to be associated with discontent about the parking situation. For the most part, the most frequent complaint of both public users and federal employees was the limited and costly parking. The parking problem has since been alleviated, in part, by the recent conversion of 15 spaces behind the building to public use.

## Interrelationships Between the Ann Arbor Federal Building and the Community

The extent to which the building has been integrated into the downtown area was determined in four ways. First, community residents contacted by telephone were asked whether or not they knew where the building was located. Second, they were asked how often they had visited the building. Third, they, along with persons contacted at the building, were asked how well they thought the building fit into downtown Ann Arbor. Finally, building occupants were asked whether they were more likely to use a number of downtown facilities and services since moving to the Federal Building.

*Do Ann Arbor residents know where the building*
*is located, and how many have been there?*

More than eight in ten adult residents contacted by telephone knew where the Ann Arbor Federal Building was located, and 75 percent said they had visited the building at one time or another (see Table 5.1). Among those who had been to the building, nearly half (45 percent) had visited it during the past month. On average, visitors had been to the building 2.9 times during the previous month.

We had expected that knowing the location of the building and using it would be associated with several specific resident characteristics. For example, it was hypothesized that persons living and working in the downtown area would be more likely to know where the building is and be more likely to use it. Similarly, we expected that students at the nearby University of Michigan campus would be more knowledgeable about the building and more inclined than other residents to use it. The findings show that although there was a tendency for people who live, work, or study in central Ann Arbor to know the whereabouts of the Federal Building and to use it, the relationships were statistically insignificant.[1] Knowledge of the building's location and its use are only associated with the number of visits to downtown Ann Arbor. The more frequently people visited downtown, the more likely they were to know where the Federal Building is located and to conduct business there.

Among the building users interviewed on-site, fewer than 10 percent (five people) said it had been their first visit to the building. The remaining users averaged 7.4 visits to the building during the past month.

Another issue related to the integration of the building into the fabric of downtown Ann Arbor is the extent to which community residents use the plaza in front of the building. More than three in ten community residents interviewed by telephone said they had used the plaza at one time or another. An identical proportion of on-site visitors reported using the plaza. For the most part, people who used the plaza said they had lunch there, met friends, or just relaxed on one of the benches.

*How well does the building fit into downtown Ann Arbor?*

Most community residents and on-site visitors said the building fit in well with downtown Ann Arbor. On-site visitors were somewhat more

TABLE 5.1

Percentage of Community Residents Knowing the Location
of Federal Building and Having Visited It

| Characteristics of Residents | Know Location of Building[a] | Have Ever Visited Building[b] | Visited Building Last Month[c] |
|---|---|---|---|
| Total Sample | 84(113)[d] | 75(113) | 45(84) |
| Residential Location | | | |
| Central Ann Arbor | 90( 10) | 70( 10) | 57( 7) |
| Elsewhere | 83(103) | 76(103) | 43(75) |
| Length of Residence in Ann Arbor | | | |
| 2 years or less | 92( 25) | 64( 25) | 44(16) |
| 3–12 years | 77( 43) | 74( 43) | 52(31) |
| Longer than 12 years | 88( 41) | 82( 41) | 42(33) |
| Student Status | | | |
| U-M Student | 90( 20) | 75( 20) | 53(15) |
| Non-Student | 81( 91) | 75( 91) | 43(65) |
| Employment Status | | | |
| Currently employed | 84( 61) | 79( 61) | 40(47) |
| Unemployed | 83( 52) | 71( 52) | 49(37) |
| Place of Employment | | | |
| Downtown Ann Arbor | 90( 10) | 80( 10) | 83( 6) |
| U-M Campus | 84( 19) | 84( 19) | 44(16) |
| Elsewhere | 81( 32) | 75( 32) | 30(23) |
| Number of Downtown Visits Last Month | | | |
| None | 45( 11) | 27( 11) | —(—) |
| 1–4 | 73( 37) | 68( 37) | 42(24) |
| 5–10 | 96( 28) | 82( 28) | 44(23) |
| 11–20 | 94( 18) | 89( 18) | 57(14) |
| More often | 100( 16) | 100( 16) | 50(16) |

[a] Knowing where the building was located is based on affirmative responses to several questions: "First of all, have you ever been in the Ann Arbor Federal Building?" and for those who said no, "Do you know where the building is?" and "Could you tell me where it is?" For respondents who did not know the location, an additional question was asked: "Are you familiar with the light-brown tiled building downtown with the Post Office in it, the one on Liberty Street?"

[b] Respondents who knew where the building was located by virtue of their response to the question, "Are you familiar with the light-brown tiled building...?" were asked: "Well that's the Ann Arbor Federal Building. Have you ever been in the building?" Affirmative responses were combined with those given to the question, "First of all, have you ever...?"

[c] For respondents who had been to the Federal Building, the question was asked: "About how many times during the past month have you been there?" The mean number of visits was 2.9 times for respondents who had visited the building.

[d] Numbers in parentheses represent the bases for percentages.

TABLE 5.2

Community Residents and On-Site Visitors Ratings
of How Well Federal Building Fits into Downtown Ann Arbor
(Percentage Distribution)

| Building Fits in: | Community Residents | On-Site Visitors |
|---|---|---|
| Very well | 29 | 35 |
| Fairly well | 43 | 46 |
| Not very well | 17 | 14 |
| Not well at all | 11 | 5 |
| Total | 100 | 100 |
| Number of respondents | 102 | 59 |
| Mean rating[a] | 2.9 | 3.1 |

[a] Mean ratings are based on scores of 4 for "very well," 3 for "fairly well," 2 for "not very well," and 1 for "not well at all."

likely to give this response than were telephone respondents (81 percent vs. 72 percent; see Table 5.2).

Among the residents contacted by telephone, no significant differences were found in ratings between those living and working in different parts of Ann Arbor. Similarly, length of residence in Ann Arbor, student status, and the number of visits to the downtown area were not associated with ratings. However, women were more likely than men to express the opinion that the building fits well into the downtown area.

*How do the building occupants use downtown Ann Arbor?*

Another indication of the extent to which the building has been successfully integrated into the fabric of downtown Ann Arbor is the degree to which building occupants use downtown facilities and services. If it were found that the building employees made greater use of such facilities and services than they had prior to moving to the building, it would appear that, from an economic perspective, the building has contributed to the vitality of downtown Ann Arbor. As Table 5.3 shows, significant numbers of federal employees said they more often engaged in downtown activities since their agency moved to the new building. For example, two-thirds of the occupants said they were more likely to conduct personal business downtown, more than half reported eating at a restaurant more often, and four in ten said they were more likely to use the public library since moving to the Federal Building. Employees of HCRS and the Social Security Administration were

TABLE 5.3

Downtown Activities Employees More Often Engage In, by Agency
(Percentage Saying "Yes")[a]

| | | | | Agency | | | | |
|---|---|---|---|---|---|---|---|---|
| Activity | All | Post Office | IRS | Military Recruiters | HCRS | Social Security | Weather Service | Small Agencies |
| Conduct personal business downtown | 66 | 57 | 62 | 58 | 84 | 81 | 56 | 46 |
| Shop downtown or in campus area | 61 | 49 | 62 | 58 | 74 | 83 | 56 | 39 |
| Eat lunch in restaurant | 56 | 37 | 56 | 54 | 73 | 65 | 50 | 57 |
| Walk at lunch time | 46 | 16 | 70 | 50 | 36 | 68 | 31 | 50 |
| Use the public library | 41 | 22 | 22 | 58 | 51 | 57 | 56 | 57 |
| Meet friends for lunch | 34 | 17 | 28 | 50 | 49 | 39 | 25 | 40 |
| Use downtown recreational facilities | 14 | 5 | 9 | 39 | 15 | 25 | 19 | 5 |
| Number of respondents[b] | 203 | 41 | 45 | 13 | 40 | 28 | 16 | 20 |

[a] The question, "Since you started working in the building are you more likely than before to (name of activity)?" was asked for each of seven activities with respondents checking either a "yes" or "no" box.

[b] Numbers represent the minimum number of responses for any item. Nonresponse was most likely to occur among postal employees.

most likely to increase their downtown activities; those least likely to be more involved were the postal workers.

It was expected that federal employees working for agencies previously located outside the downtown area would be more likely to have increased their use of downtown facilities than employees in agencies previously situated in the downtown area. We found that, while this pattern was certainly true with respect to the three above-mentioned agencies, employees of other federal offices varied in their relative use of the downtown area. For instance, Internal Revenue Service and Weather Service personnel, who had previously worked outside of central Ann Arbor, were less likely to use downtown facilities than the military recruiters, a small group who previously worked in rented space near the new structure. These comparisons can be seen in Table 5.4, which presents a summary measure, or index, of the comparative downtown use by employees in the different agencies.[2] Clearly, there are factors other than the agencies' prior locations that are associated with the employees' comparative use of downtown Ann Arbor.

Using a series of multiple classification analyses, we were able to examine which of several different sets of employee characteristics had the greatest influence on whether or not they used facilities and services in downtown Ann Arbor more often since moving to the Federal Building.[3] As Table 5.5 shows, four characteristics taken together accounted for nearly one-fifth of the variation in scores on the index for comparative use of the downtown. The most important single predictor was the sex of the employee; women used downtown facilities and services with greater regularity since their agency had moved to the Federal Building.

The second most important factor was how employees traveled to and from work. Those who walked, biked, or rode a bus were much more likely to use downtown facilities than employees who drove to work. And federal employees who had worked in the building for less than six months at the time of our survey were less inclined to use the downtown more frequently.

*How do the building occupants*
*evaluate the location of the building?*

The success of the site decision for the building — to place it on Liberty Street in downtown Ann Arbor — can be judged in part by the ratings given to its location by the occupants. For the most part, the federal employees were quite satisfied with the building's location. Three-fourths rated the location as either excellent or pretty good, while only 10 percent said it was poorly situated (see Table 5.6). Highest ratings were given by the military recruiters, employees of the Social Security

TABLE 5.4

Comparative Use of Downtown Ann Arbor, by Agency
(Percentage Distribution)

| Comparative Downtown Use Index[a] | All | Agency | | | | | | |
| | | Post Office | IRS | Military Recruiters | HCRS | Social Security | Weather Service | Small Agencies |
| --- | --- | --- | --- | --- | --- | --- | --- | --- |
| High (5) | 17 | 17 | 8 | 23 | 15 | 31 | 12 | 20 |
| (4) | 7 | 2 | 2 | – | 13 | 15 | 6 | 12 |
| (3) | 14 | 8 | 17 | 38 | 21 | 6 | 13 | 8 |
| (2) | 26 | 12 | 44 | – | 32 | 27 | 25 | 20 |
| Low (1) | 36 | 61 | 29 | 39 | 19 | 21 | 44 | 40 |
| Total | 100 | 100 | 100 | 100 | 100 | 100 | 100 | 100 |
| Number of respondents | 233 | 54 | 49 | 13 | 47 | 33 | 16 | 27 |
| Mean score | 2.4 | 2.0 | 2.2 | 2.7 | 2.7 | 3.1 | 2.2 | 2.3 |

[a] For each individual, an index score was created by summing the number of affirmative responses to the question, "Since you started working in the building are you more likely than before to: ____?" The seven activities asked about were: eat lunch in a restaurant, go for a walk at lunch time, shop downtown or in the campus area, meet friends for lunch, use the public library, use downtown recreational facilities, and conduct personal business. The higher the score, the greater the likelihood that an individual uses the downtown area more than before.

TABLE 5.5

Comparative Downtown Use Predicted
by Selected Employee Characteristics
(Multiple Classification Analysis; N = 225)

| Predictor | Eta Coefficient | Beta Coefficient | Adjusted Use Scores |
|---|---|---|---|
| Sex of employee | .39 | .36[a] | Women (3.09); men (2.05) |
| Mode of travel | .28 | .25 | Walkers, bikers (3.25); bus riders (3.07); drivers (2.21) |
| Time in building | .16 | .17 | Less than 6 months (1.89) |
| Agency | .19 | .16 | Military recruiters (3.17); postal workers (2.28); IRS (2.30) |
| Percentage of variance explained (adjusted multiple $R^2$) | | 19.2 | |
| Overall Use Score: 2.44[b] | | | |

[a] The predictors are listed in order of importance.

[b] Scores ranged from 5 for a high use of downtown to 1 for low use.

Administration, and those in the smaller agencies, while the Weather Service staff and the postal workers rated the location less favorably.

Several factors were believed to be associated with the employees' ratings of the building location. First, it was hypothesized that people who became more active users of the downtown area as a result of the move would give higher marks to the location than those individuals who had worked in downtown Ann Arbor before the move. Second, we expected that employees who no longer drove to work would rate the location most favorably, while those who had previously driven and still drove would rate the location poorly. Finally, we expected that employees who recently had begun working in the building would be more likely to explore the downtown during their lunch period or after work and consequently would find the location attractive. In order to test these hypotheses and, if correct, see whether the relationships were maintained once other factors were accounted for, we looked at how well several factors—comparative downtown use, change in travel mode, length of time in the building, and agency—predicted people's ratings of the location. As Table 5.7 shows, these predictors taken together explained 18.6 percent of the variance in the ratings. Although each predictor was to some degree related to employee ratings of the downtown location, comparative downtown use was the most important. That is, once the other factors were accounted for, the extent to which employees were more or less likely to use the downtown was

TABLE 5.6

Rating of Building Location, by Agency[a]

| Rating of Location | All | Post Office | IRS | Military Recruiters | HCRS | Social Security | Weather Service | Small Agencies |
|---|---|---|---|---|---|---|---|---|
| | | | | Agency | | | | |
| Excellent | 35 | 34 | 28 | 54 | 43 | 39 | 19 | 30 |
| Pretty good | 39 | 30 | 45 | 31 | 30 | 46 | 38 | 55 |
| Fair | 16 | 21 | 17 | 7 | 19 | 9 | 12 | 15 |
| Poor | 10 | 15 | 10 | 8 | 8 | 6 | 31 | — |
| Total | 100 | 100 | 100 | 100 | 100 | 100 | 100 | 100 |
| Number of respondents | 236 | 53 | 47 | 13 | 47 | 33 | 16 | 27 |
| Mean ratings[b] | 3.0 | 2.8 | 2.9 | 3.3 | 3.1 | 3.2 | 2.7 | 3.2 |

[a] The question was: "How would you rate the location of the Federal Building as a place to work?"

[b] Mean ratings are based on scores of 4 for "excellent," 3 for "pretty good," 2 for "fair," and 1 for "poor."

TABLE 5.7

Rating of Building Location Predicted
by Selected Employee Characteristics
(Multiple Classification Analysis; N = 228)

| Predictors | Eta Coefficient | Beta Coefficient | Adjusted Ratings |
|---|---|---|---|
| Comparative downtown use | .36 | .42[a] | Low use (2.55); high use (3.13) |
| Time in building | .11 | .22 | More than 2 years (2.77); less than 6 months (3.26) |
| Agency | .16 | .21 | Small agencies (3.31); Weather Service (2.42) |
| Change in travel mode | .16 | .15 | Car to car pool (2.79); car to bus, walk, bicycling (3.13); no change — bus, walk (3.12) |
| Percentage of variance explained (adjusted multiple $R^2$) | | 18.6 | |
| Overall Rating: 2.97[b] | | | |

[a] The predictors are listed in order of importance.

[b] Ratings are based on scores of 4 for "excellent," 3 for "very good," 2 for "fair" and 1 for "poor."

most strongly associated with their feelings about the building's location.

It should also be noted that the hypothesis concerning change in travel mode was only partially correct. While employees who no longer drove their automobiles as a result of the move rated the location more favorably than did employees who drove before and still drove, they were no more positive in their evaluations than employees who had previously taken a bus, walked, or biked to work and who currently traveled in a similar manner.

## How does the building location compare to employees' previous work locations?

Employees' ratings of the comparative advantages of the downtown location were reflected in their responses to the question, "Compared to where you worked before, is the location of the Federal Building better, worse, or the same?" (Figure 5.1). Among all employees, positive responses outnumbered negative responses by two to one (43 percent versus 22 percent). Feelings about the comparative advantages of the new location were strongest among the military recruiters and the postal workers, although nearly a third from each agency said the new location was no better or worse (see Table 5.8). The majority of workers in the small agencies and in IRS also felt the location offered

FIGURE 5.1

Comparative Assessment of Building's Location —
Current Situation Relative to Past Situation
(Difference between respondents saying location
is better and respondents saying it is worse.)

no real advantages over their previous locations. On the other hand, a
clear majority in all other agencies except one indicated the new location was more favorable. Most Weather Service employees said the
location of the building was worse than their previous location.

Not surprisingly, employees said the advantages of the downtown
location are the building's proximity to stores, restaurants, banks, and
other amenities. These feelings were most strongly expressed by the
staff of the Social Security Administration, the majority of whom were
women. It should be remembered that in our analysis of factors associated with the relative use of downtown, the employee's sex was the
most important predictor, even after the employee's agency was taken
into account.

The major disadvantage of the downtown location seems to be the
parking situation. When employees who said the location was worse
were asked to indicate why, the most frequently mentioned reasons
were related to parking. The inconvenience or lack of parking or its
high cost were major complaints expressed by people in every agency
(see Table 5.8).

TABLE 5.8

Comparative Evaluation of Building Location, by Agency
(Percentage Distribution)

| Evaluations | All | Agency | | | | | | |
| --- | --- | --- | --- | --- | --- | --- | --- | --- |
| | | Post Office | IRS | Military Recruiters | HCRS | Social Security | Weather Service | Small Agencies |
| Comparative Location Evaluation[a] | | | | | | | | |
| Better | 43 | 51 | 25 | 54 | 54 | 50 | 33 | 32 |
| Worse | 22 | 17 | 23 | 8 | 22 | 20 | 53 | 24 |
| Same | 27 | 30 | 40 | 30 | 13 | 17 | 7 | 44 |
| Better and worse | 8 | 2 | 12 | 8 | 11 | 13 | 7 | — |
| Total | 100 | 100 | 100 | 100 | 100 | 100 | 100 | 100 |
| Number of respondents | 230 | 53 | 48 | 13 | 46 | 30 | 15 | 25 |
| Reasons Location is Better | | | | | | | | |
| Near stores, restaurants, banks | 33 | 13 | 55 | — | 41 | 63 | 17 | 14 |
| Central location; near downtown, campus | 25 | 32 | 10 | 45 | 27 | 16 | 25 | 14 |
| Close to home | 12 | 7 | 10 | 9 | 15 | 11 | 17 | 29 |
| Convenient for public, proximity to service area | 2 | — | 5 | 19 | — | — | — | — |
| Building is better, newer, cleaner | 6 | 16 | — | — | 2 | 5 | 8 | — |
| Public transportation is close, convenient | 5 | 7 | — | 9 | 8 | — | 8 | — |
| Parking, traffic is bad | 2 | 3 | — | 9 | 2 | — | — | — |
| Other | 15 | 22 | 20 | 9 | 5 | 5 | 25 | 43 |
| Total | 100 | 100 | 100 | 100 | 100 | 100 | 100 | 100 |
| Total mentions | 141 | 31 | 20 | 11 | 41 | 19 | 7 | 12 |
| Number of respondents | 116 | 28 | 17 | 8 | 30 | 19 | 6 | 8 |

TABLE 5.8 (Continued)

| Evaluations | All | Agency | | | | | | |
| --- | --- | --- | --- | --- | --- | --- | --- | --- |
| | | Post Office | IRS | Military Recruiters | HCRS | Social Security | Weather Service | Small Agencies |
| Reasons Location is Worse | | | | | | | | |
| Parking is inconvenient, unavailable | 26 | 31 | 19 | 50 | 32 | 31 | – | 33 |
| Traffic congestion, one-way streets | 21 | 62 | 15 | – | 17 | 8 | 56 | – |
| Parking is costly | 21 | – | 25 | – | 17 | 46 | 22 | 20 |
| Far from home | 4 | – | 4 | – | 3 | – | – | 13 |
| Building is worse, poor design | 2 | – | 4 | – | 3 | – | – | – |
| Inconvenient for public, farther from area of operations | 5 | – | 8 | 50 | – | – | – | 7 |
| Other | 21 | 7 | 25 | – | 28 | 15 | 22 | 27 |
| Total | 100 | 100 | 100 | 100 | 100 | 100 | 100 | 100 |
| Total mentions | 110 | 13 | 27 | 4 | 29 | 13 | 15 | 9 |
| Number of respondents | 68 | 9 | 17 | 2 | 15 | 10 | 9 | 6 |

a The question was: "Compared to where you worked before, is the location of the Federal Building better, worse, or the same?" Of the 239 who completed the questionnaire 5 did not answer the question and 4 said they were not previously employed.

TABLE 5.9

Mode of Travel to the Federal Building among Visitors
(Percentage Distribution)

| Mode of Travel | Community Residents | On-Site Visitors |
|---|---|---|
| Drive | 62 | 45 |
| Walk | 32 | 38 |
| Bus | 1 | 11 |
| Other | 5 | 6 |
| Total | 100 | 100 |
| Number of respondents | 84 | 55 |

## Travel to the Federal Building

Questions dealing with transportation to and from the building centered on how community residents and federal employees traveled there, whether the employees' modes of travel had changed as a result of the move from their prior locations, and the extent to which federal employees felt the trip to and from work was convenient.

*What means of transportation do Ann Arbor residents use when visiting the Federal Building?*

Among the Ann Arbor residents contacted by telephone, somewhat more than 60 percent said they most often drove when visiting the building (see Table 5.9); one-third said they walked, two people said they biked, and only one person said he usually came by bus. Among visitors contacted at the building itself, a somewhat smaller proportion (45 percent) said they drove, 38 percent indicated they were most likely to walk, and, surprisingly, one-tenth of the visitors said they came by bus.[4] When asked how they had arrived at the Federal Building on that particular trip, most people gave the identical response, suggesting that people who visit the building regularly do not vary from trip to trip in the means of getting there.

*What means of transportation do the federal employees use when going to and from work?*

As Table 5.10 shows, more than eight in ten employees came to work by car: 61 percent drove their own car, 17 percent said they participated in a car pool, and 4 percent drove a car belonging to a federal agency. Another 10 percent indicated they usually came by bus and 7 percent said they most often walked. There were clear differences in

TABLE 5.10

Travel to Work, by Agency
(Percentage Distribution)

| Travel Behavior | All | Agency | | | | | | |
| --- | --- | --- | --- | --- | --- | --- | --- | --- |
| | | Post Office | IRS | Military Recruiters | HCRS | Social Security | Weather Service | Small Agencies |
| Current Mode of Travel[a] | | | | | | | | |
| Own car | 61 | 70 | 69 | 54 | 36 | 52 | 94 | 65 |
| Agency car | 4 | – | – | 38 | – | – | – | 19 |
| Car pool[b] | 17 | 20 | 25 | – | 15 | 24 | – | 12 |
| Bus | 10 | 2 | 6 | 8 | 26 | 15 | 6 | – |
| Walk | 7 | 6 | – | – | 21 | 6 | – | 4 |
| Other[c] | 1 | 2 | – | – | 2 | 3 | – | – |
| Total | 100 | 100 | 100 | 100 | 100 | 100 | 100 | 100 |
| Number of respondents | 238 | 54 | 49 | 13 | 47 | 33 | 16 | 26 |
| Previous Mode of Travel[d] | | | | | | | | |
| Own car | 83 | 90 | 92 | 84 | 69 | 84 | 88 | 73 |
| Agency car | 2 | – | – | – | – | – | – | 19 |
| Car pool | 7 | 4 | 4 | 8 | 13 | 10 | – | 4 |
| Bus | 3 | – | 2 | 8 | 5 | 6 | 6 | 4 |
| Walk | 4 | 4 | 2 | – | 11 | – | 6 | – |
| Other | 1 | 2 | – | – | 2 | – | 6 | – |
| Total | 100 | 100 | 100 | 100 | 100 | 100 | 100 | 100 |
| Number of respondents | 233 | 54 | 48 | 13 | 45 | 31 | 16 | 26 |

| | | | | | | | | |
|---|---|---|---|---|---|---|---|---|
| **Change in Travel Mode** | | | | | | | | |
| No change – car to car | 66 | 70 | 68 | 92 | 36 | 55 | 93 | 88 |
| Car to car pool | 12 | 17 | 21 | – | 5 | 17 | – | 4 |
| Car to bus, walk, other | 11 | 5 | 4 | – | 32 | 17 | – | 4 |
| No change – bus, walk, other | 11 | 8 | 7 | 8 | 27 | 11 | 7 | 4 |
| Total | 100 | 100 | 100 | 100 | 100 | 100 | 100 | 100 |
| Number of respondents[e] | 226 | 53 | 47 | 13 | 44 | 29 | 25 | 15 |
| **Travel Time to Work** | | | | | | | | |
| Less than 15 minutes | 29 | 37 | 12 | 25 | 26 | 43 | 37 | 30 |
| 15–29 minutes | 42 | 41 | 37 | 33 | 57 | 30 | 31 | 48 |
| 30–44 minutes | 17 | 20 | 31 | 17 | 9 | 12 | 19 | 7 |
| 45–59 minutes | 11 | 2 | 20 | 25 | 4 | 15 | 13 | 11 |
| One hour or more | 1 | – | – | – | 4 | – | – | 4 |
| Total | 100 | 100 | 100 | 100 | 100 | 100 | 100 | 100 |
| Number of respondents | 238 | 54 | 49 | 12 | 47 | 33 | 16 | 27 |
| Mean travel time (minutes) | 24 | 20 | 31 | 29 | 23 | 22 | 24 | 24 |

[a] The question was: "How do you usually get to work?"

[b] Of the employees who said they shared a ride or came to work as part of a car pool, three said they never drove themselves.

[c] Other includes bicycling.

[d] The question was: "Before you began working in this building, how did you usually get to and from work?"

[e] Miscellaneous changes in travel mode were made by seven additional employees. These included such changes as bus to car, walk to car pool, and bicycle to agency car.

*Public transportation is conveniently located for users of the Federal Building. One out of every ten employees came to work by bus, and a similar proportion of visitors contacted at the building said they used public transportation.*

the mode of transportation to the building for people working in the different agencies. Virtually everyone in the Weather Service drove his or her own car. In contrast, driving was reported by only a third of the HCRS employees and somewhat more than half of the military recruiters and Social Security workers. Employees in only two agencies, the military recruiting offices and the FBI, indicated they use an agency car.[5] It is interesting to note that reports of public transportation use and walking were most prevalent among employees of HCRS. One quarter of the people in that agency rode buses, while a somewhat smaller proportion said they usually walked to work.

*Has the mode of travel changed for the federal employees since moving to the new building?*

As a way of determining whether travel habits had changed, employees were asked, "Before you began working in this building, how did you usually get to and from work?" The automobile clearly dominated the mode of travel for federal employees prior to their move downtown. While six in ten respondents said they currently drove their own car, eight in ten indicated that they had driven to work before the move. By examining the questions about current and past modes of travel in relation to each other, it was possible to see the exact nature of the changes made in the journey to work. The third part of Table 5.10 shows that about one-quarter of the federal employees gave up driving to work in order to carpool, walk, bicycle, or take a bus. The mode of travel did not change for the remaining 77 percent after moving to the new building.

Changes to public transportation and walking were most prevalent among HCRS personnel and those working in the Social Security Administration. Least likely to change their mode of travel were employees of the Weather Service and the military recruiters; more than nine out of ten employees in those agencies drove both before and after the move to downtown, while the remainder did not change their pattern of walking or riding a bus or bicycle.

*How long does it take the federal employees to get to work?*

To an extent, travel time is related to mode of transportation. For instance, IRS employees and the military recruiters, whose average travel time was approximately a half hour, were among those more likely to drive to work. Weather Service personnel, who were also heavy automobile users, spent an average of 24 minutes travelling to work. Most likely to ride a bus, bicycle, or walk were the HCRS and Social Security employees, who spent an average of about 22 minutes in the work trip. While travel time may be interesting in and of itself, it has greater meaning when viewed in light of how employees feel about their work trip. Indeed, three out of every four employees who spent less than 15 minutes on their work trip said that travel to and from work was convenient. In contrast, this response was given by just 38 percent of those travelling 15 to 29 minutes, 22 percent of those travelling 30 to 44 minutes, and only 14 percent of the people who spent 45 minutes or more in their work trip.

*How do building occupants rate the
convenience of travel to work?*

In response to the question about whether transportation to and from work was convenient, eight in ten of the federal employees responded positively. As seen in Table 5.11, employees in the Social Security Administration were most likely to view the work trip as convenient, and those in the Weather Service were least likely to feel this way.[6]

Efforts were made as a part of this study to understand what factors were most likely to influence people's views on the convenience of travelling to and from work. It was hypothesized that, in addition to travel time, the current mode of travel and changes in mode would be associated with people's ratings. Those employees who walked, rode a bus, or biked were expected to be more likely to view travel as convenient compared with people who came by automobile. Similarly, it was expected that after the move to the new building, people who changed their mode from driving to either using public transit, walking, or biking would rate convenience more positively than those who had not changed their travel mode. Both hypotheses were supported by the data (Table 5.12). Employees who walked or bicycled tended to give higher ratings than those who took a bus, while the lowest ratings were reported by employees who drove or shared a ride with someone else. At the same time, the most positive ratings of convenience were found among people who changed their mode from the automobile to riding the bus, bicycling, or walking.

The question of whether mode of travel was more important in people's ratings of the convenience of their work trip was raised and examined in a multivariate analysis. As Table 5.13 shows, travel time and current mode of travel both can be used to predict people's ratings of travel convenience. Of the two factors, however, the length of the work trip was found to be significantly more important than how people get there. Even after taking travel mode into consideration, travel time was shown to be the best predictor of how people feel about the convenience of the work trip. Those who spent less than 15 minutes getting to work and walked or rode a bike gave the highest ratings, while those who traveled 45 minutes or more and drove gave the lowest ratings.

TABLE 5.11

Employees' Rating of Travel Convenience, by Agency
(Percentage Distribution)

| | | | | | Agency | | | | |
|---|---|---|---|---|---|---|---|---|---|
| "Travel to and from work is convenient." | All | Post Office | IRS | Military Recruiters | HCRS | Social Security | Weather Service | Small Agencies |
| Very true | 42 | 43 | 33 | 50 | 49 | 49 | 38 | 39 |
| Somewhat true | 37 | 30 | 45 | 34 | 29 | 42 | 37 | 46 |
| Not very true | 14 | 25 | 8 | 8 | 9 | 9 | 19 | 15 |
| Not at all true | 7 | 2 | 14 | 8 | 13 | – | 6 | – |
| Total | 100 | 100 | 100 | 100 | 100 | 100 | 100 | 100 |
| Number of respondents | 234 | 53 | 49 | 12 | 45 | 33 | 16 | 26 |

TABLE 5.12

Ratings of Travel Convenience to Work,
by Current Travel Mode and Change in Travel Mode
(Percentage Distribution)

| | "Travel To and From Work is Convenient" | | | | | |
| Mode of Travel | Very True | Somewhat True | Not Very True | Not True at All | Total | Number of Respondents |
| --- | --- | --- | --- | --- | --- | --- |
| Current Mode | | | | | | |
| Own/agency car | 37 | 40 | 15 | 8 | 100 | 151 |
| Car pool | 25 | 46 | 22 | 7 | 100 | 41 |
| Bus | 64 | 32 | — | 4 | 100 | 22 |
| Walk, other | 95 | 5 | — | — | 100 | 19 |
| Change in Mode | | | | | | |
| No change — car | 37 | 39 | 16 | 8 | 100 | 145 |
| Car to car pool | 22 | 56 | 19 | 3 | 100 | 27 |
| No change — bus, walk, other | 60 | 20 | 12 | 8 | 100 | 25 |
| Car to bus, walk, other | 75 | 21 | — | 4 | 100 | 24 |

TABLE 5.13

Rating of Travel Convenience Predicted by Travel Characteristics
(Multiple Classification Analysis; N = 226)

| Predictors | Eta Coefficient | Beta Coefficient | Adjusted Ratings |
| --- | --- | --- | --- |
| Travel time to work | .47 | .44[a] | Less than 15 minutes (1.38); 45 minutes or more (2.67) |
| Mode of travel | .29 | .29 | Car (1.96), walk, bike (1.21), bus (1.35) |
| Agency | .00 | .22 | Small agencies (1.44); HCRS (2.14) |
| Percentage of variance explained (adjusted multiple $R^2$) | | 26.7 | |
| Overall Rating: 1.85[b] | | | |

[a] The predictors are listed in order of importance.

[b] Ratings to the statement, "Travel to and from work is convenient," are 1 for "very true," 2 for "somewhat true," 3 for "not very true," and 4 for "not at all true."

*Despite its central location and the popularity of bicycling in Ann Arbor, only one in twenty visitors rode a bicycle to the Federal Building. Among federal workers, only two said they most often came to work by bicycle.*

*Access to the short-term parking lot along Fifth Avenue was a source of traffic congestion and annoyance to building users. The lot contains 15 spaces, including one for the handicapped.*

## Parking

Despite its central location and its proximity to public transportation, most federal employees and visitors to the Ann Arbor Federal Building arrived by car. As part of our evaluation, the question of where these drivers were likely to park was addressed, and a systematic examination was made of the parking situation as viewed by the drivers.[7] Earlier we had been told that parking was indeed a problem, and competition for parking space was rampant between agency personnel and the public. Forty parking spaces for employees were provided behind the building, and the postal workers had access to leased space in a lot to the west on Fourth Avenue, but most federal employees who drove used paid public parking facilities available to the east and west of the building. Some even resorted to metered street parking along Liberty and on side streets.

Community residents visiting the building can use the short-term parking lot along Fifth Avenue. This lot contains 15 spaces, including one designated for handicapped persons, and is limited to a 15-minute stay for Post Office patrons. Our research team found the lot to be filled at most times during the day.

*Where do community residents who drive usually park?*

Half of the community residents interviewed by telephone said that when visiting the Federal Building they most often used the short-term

TABLE 5.14

Parking Places and Problems, as Reported by
Community Residents and On-Site Visitors
(Percentage Distribution)

| Parking | Community Residents | On-Site Visitors |
|---|---|---|
| Parking Place | | |
| Short-term lot | 46 | 54 |
| Street | 21 | 15 |
| City lot on 5th Street | 19 | 8 |
| City parking structure | 8 | 15 |
| Elsewhere | 6 | 8 |
| Total | 100 | 100 |
| "Have you ever had problems with the parking there?" | | |
| Percentage saying "yes" | 58 | 46 |
| Convenience of Parking[a] | | |
| Very convenient | 16 | 12 |
| Fairly convenient | 31 | 27 |
| Not very convenient | 39 | 42 |
| Not at all convenient | 14 | 19 |
| Total | 100 | 100 |
| Number of respondents | 51 | 26 |

[a] "In general, how convenient is parking (around here)? Would you say it's *very* convenient, *fairly* convenient, *not very* convenient, or *not at all* convenient?"

parking lot next to the Post Office; one in five said they parked in the municipal lot behind the library, and another one in five said they parked on the street (see Table 5.14). Among patrons interviewed at the building, about half of those who drove were using the Post Office lot that day (54 percent), while a somewhat smaller group had parked in the library lot (8 percent) or on the street (15 percent).

Both groups of outside users indicated they had problems when driving to the building; 58 percent contacted by telephone reported having difficulties, while 46 percent interviewed at the building had problems with parking that very day. For the most part, these problems related to access to the short-term lot and the fact that space was not always available. If more than two cars were waiting, traffic along Fifth Avenue would be blocked. Most often mentioned by both groups, however, was a general lack of parking facilities. Indeed, in response to the question "How convenient is the parking?" 53 percent of the drivers contacted by telephone said parking near the Federal Building was not at all convenient, and 61 percent of the on-site users gave this response (Table 5.14).

*Half of the community residents and six in ten federal employees who drove to the building reported parking problems.*

*The people who mentioned problems with parking most frequently cited the limited number of nearby spaces and illegally parked cars.*

It is interesting to note that the parking problems faced by the public were readily recognized by federal employees. One Social Security employee wrote:

> Our office services the entire county, so we have many people who come to us by car. They sometimes park in the small Post Office lot, not realizing they have to be in our office for several hours. They also don't realize that there is a $25 parking ticket issued for over-parking. Or they may park at one of the meters in the parking structure across the street or at the library. Here the parking tickets are only $3. They usually underestimate the time they'll be in our office, so they frequently get up in the middle of the interview to run out to feed the meters. This is very disruptive to the interview procedure, and understandably makes our staff and the public angry. Many of these are poor people who cannot afford parking tickets. And many refuse to return to our office because of the parking problems. And I have not even touched on my own feelings about not having a place to park as an employee unless I spend $25 each month, a loss in pay I cannot afford.

## Where do federal employees usually park when they drive to work?

Employees who drove to work clearly did not park in any single location. With the exception of the postal workers who leased space on Fourth Avenue, most drivers parked in either the city parking structure (31 percent) or in the lot behind the building (29 percent). Another 14 percent said that they most often parked on the street, while the remainder who drove parked elsewhere in the downtown area. Most likely to use the city parking structure were employees of IRS, Social Security Administration, the Weather Bureau, HCRS, and the small agencies. In addition to selected individuals of each of these agencies, agents from the FBI and employees from the Soil Conservation Service, the Defense Department, and the military recruiters used the parking lot behind the building.

## How do employees who drive feel about parking?

The employees at the Ann Arbor Federal Building who drove to work gave poor ratings to parking facilities. Most had had free parking available to them at their agencies' former locations. For example, several agencies were previously located in rented office space in either a shopping center (IRS), a suburban office building (Social Security Administration), or an industrial park (HCRS). The Weather Bureau offices were at the Detroit Metropolitan Airport. At each of these locations, free parking was available in considerable abundance.

As the second part of Table 5.15 shows, about half of the 178 employees who drove said parking was less convenient than where they

TABLE 5.15

Parking Places and Problems, by Agency
(Percentage Distribution)

| Parking | All | Agency | | | | | | |
| --- | --- | --- | --- | --- | --- | --- | --- | --- |
| | | Post Office | IRS | Military Recruiters | HCRS | Social Security | Weather Service | Small Agencies |
| Parking Place | | | | | | | | |
| Street | 14 | 9 | 7 | – | 46 | 5 | 47 | – |
| City parking structure | 31 | – | 54 | 8 | 33 | 60 | 40 | 24 |
| Lot behind building | 29 | 17 | 35 | 92 | – | 20 | 13 | 60 |
| Post office lot on 4th Street | 20 | 70 | 2 | – | 17 | 5 | – | – |
| City parking lot on 5th Street | 3 | 4 | – | – | 4 | – | – | 8 |
| Elsewhere | 3 | – | 2 | – | – | 10 | – | 8 |
| Total | 100 | 100 | 100 | 100 | 100 | 100 | 100 | 100 |
| Number of respondents | 192 | 47 | 46 | 12 | 24 | 20 | 15 | 25 |
| Convenience of Parking[a] | | | | | | | | |
| More convenient | 24 | 50 | 17 | 23 | 4 | 10 | 7 | 30 |
| Less convenient | 48 | 20 | 50 | 31 | 83 | 75 | 72 | 40 |
| About the same | 28 | 30 | 33 | 46 | 13 | 15 | 21 | 30 |
| Total | 100 | 100 | 100 | 100 | 100 | 100 | 100 | 100 |
| Number of respondents | 178 | 46 | 42 | 12 | 23 | 20 | 14 | 20 |
| "Have you ever had problems with parking?" | | | | | | | | |
| Percentage saying "yes" | 59 | 83 | 41 | 46 | 55 | 58 | 68 | 50 |
| Number of respondents[b] | 175 | 42 | 44 | 13 | 22 | 19 | 15 | 20 |

TABLE 5.15 (Continued)

| | | | | | Agency | | | |
|---|---|---|---|---|---|---|---|---|
| Parking | All | Post Office | IRS | Military Recruiters | HCRS | Social Security | Weather Service | Small Agencies |
| Type of Parking Problem | | | | | | | | |
| Lot small, crowded, congested | 14 | 22 | 15 | 13 | 9 | – | – | – |
| Expensive parking | 12 | 4 | 15 | 13 | – | 39 | 50 | – |
| Parking tickets | 9 | 2 | 15 | – | 9 | 23 | 12 | 20 |
| Unauthorized users of reserved spaces | 13 | 14 | 5 | 12 | – | 15 | 12 | 30 |
| Not enough spaces—general mention | 34 | 47 | 20 | 50 | 46 | 8 | 13 | 30 |
| Other | 18 | 11 | 30 | 12 | 36 | 15 | 13 | 20 |
| Total | 100 | 100 | 100 | 100 | 100 | 100 | 100 | 100 |
| Number of mentions | 125 | 55 | 20 | 8 | 11 | 13 | 8 | 10 |

[a] The question was: "Compared to where you parked before you worked in the Federal Building, is your current parking more convenient, less convenient, or about the same?" Two of the 180 respondents who answered the question said that they didn't drive before or weren't previously employed.

[b] Of the employees who said they shared a ride or came to work as part of a car pool, three said they never drove themselves.

FIGURE 5.2

Comparative Assessment of Parking—
Current Situation Relative to Past Situation
(Difference between respondents saying more
convenient and respondents saying less convenient.)

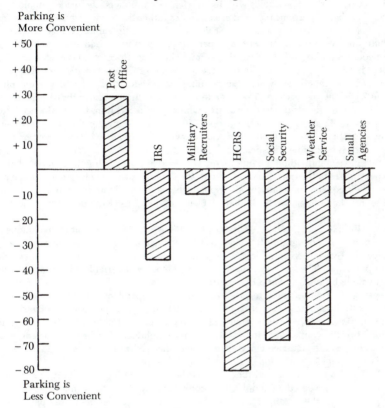

had previously worked. Only the postal workers indicated that parking was now more convenient for them. Figure 5.2 shows the comparative assessments of parking near the Ann Arbor Federal Building.

About six in ten of the employees who drove to work said they had experienced problems in parking at the Federal Building. Most likely to report problems were the postal workers who said there weren't enough spaces in their lot and that it was used by people not working for the Post Office. Other complaints touched on the high cost of parking, the fact that unauthorized users were occupying pre-assigned spaces behind the building, and the frequency with which parking tickets were being issued at expired meters.

## Notes

1. The statistical significance of differences in proportions and the precision of estimates of proportions can only be discussed in connection with the data covering the sample of community residents. Because the data are based on a sample of the total population of Ann Arbor households with listed telephone numbers, the reader should bear in mind that the proportions reported are estimates rather than exact figures. Furthermore, the precision of these estimates can be calculated because the sample was selected by probability methods. In the case of the 84 percent of community residents who knew where the Federal Building was located, this estimate has a confidence interval of nearly 7 percent. That is, if repeated samples of residents were taken, 95 out of 100 times the proportion who said they knew where the building was located would be 84 percent plus or minus 7 percent, or between 77 percent and 91 percent. In the case of the 81 percent (or 32 people) employed outside the central area who knew the location, the confidence interval is 14 percent. Since 81 percent plus 14 percentage points is greater than the 90 percent of knowledgeable respondents employed downtown, the difference between 84 percent and 90 percent is said to be statistically insignificant.

2. For each individual, an index score was created by summing the number of affirmative responses to the question dealing with their comparative involvement in seven activities. The frequency of scores for all respondents was divided into five groups. The higher the score, the greater the likelihood that an individual used the downtown more than before.

3. Multiple Classification Analysis (MCA) is the principal multivariate technique used throughout this monograph. It is used to examine the relationship between each of a set of independent variables and a dependent variable while holding constant the effects of all other predictors. The statistics show how each independent variable relates to the dependent variable by means of the Eta coefficient. The analysis also shows how strongly the independent variables taken together related to the dependent variable by means of the multiple R, the square of which expresses the relationship as the percentage of variance explained. Finally, the analysis supplies, for each predictor variable, a Beta coefficient indicating its relative importance in the total variance explained. The Beta coefficient squared is an estimate of the independent contribution of the predictor with respect to the multiple $R^2$. For a complete discussion of the technique, see Andrews et al. (1973).

4. The reader is reminded that because the data covering community residents and on-site visitors are based on samples, the proportions reported are estimates. Furthermore, the precision of these estimates cannot be determined for the quota sample of on-site visitors. However, a confidence interval for the proportion of community residents responding in a certain manner can be calculated. In the case of the 62 percent of the community residents who drove, this estimate has a 10-percent confidence interval. That is, if repeated samples of residents were taken, 95 out of 100 times the proportion who would report driving to the building would fall between 52 percent and 72 percent.

5. The FBI office in the building, with seven employees, was the largest of the small agencies.

6. Eighty percent of the workers contacted in a 1977 national quality of employment survey also said that travel to and from work was very convenient or somewhat convenient. However, 52 percent gave the most positive reply, compared to 43 percent of the Federal Building employees. It should also be noted that in repeated national surveys, the proportion of employees indicating that travel was very convenient has steadily declined between 1969 and 1977 (Quinn and Staines, 1979).

7. The locations of nearby parking opportunities for people driving to the Federal Building are shown in Figure 3.3.

# 6

# Uses and Evaluations of the Building

## Overview

Two years after it opened, most Ann Arbor residents were quite familiar with the Federal Building. Eight in ten knew where the building was located and three-quarters had been there at one time or another. In this chapter, we examine how and when these visitors and federal employees enter and leave the building during the work day. We then consider their specific destinations within the building and any difficulties they might have in finding these destinations. We also examine the degree to which federal workers use building facilities and factors that contribute to their active or limited use. Finally, in order to determine how the building is viewed by the public and the people who use it, ratings on architectural quality are considered for community residents and federal employees, along with employee assessments of a number of specific building attributes.

Our findings show that public use of the agencies in the Federal Building varied greatly. People visited the Post Office, military recruiters, and the Social Security Administration with considerable regularity, but small agencies such as the FBI and the Soil Conservation Service were rarely visited. For the most part, the federal employees did not have much contact with agencies other than their own or the Post Office, and few used the conference room or lounge outside the snack bar. Most, however, had visited the snack bar during the preceding month.

From the point of view of community residents, the Federal Building is worthy of its many architectural awards; three out of four found the building attractive and based their opinions on the exterior design and the open space in front of the building. The building occupants, on the other hand, were less likely to feel the architectural awards were justified; about half gave the building low marks as a place to work, and a significant number of employees gave its architectural quality unfavorable ratings. To a large extent, feelings about architectural quality reflected employee sentiments about the general ambience of the agencies with which they were employed.

**Building Uses**

Earlier, we noted that a key design feature was the northern orientation of the building toward Liberty Street, the main pedestrian and vehicular route between the U-M campus and downtown Ann Arbor. As part of that decision, three entrances were planned along the northern facade — a main building entrance and two associated with the Post Office. A fourth entrance was introduced along the southern face to accommodate federal employees who either parked in the adjacent lot or used public transportation in coming to work; the southern entrance is ramped and can also be used by the public.

Systematic counts of people entering and leaving the building were taken at each entrance.[1] As Figure 6.1 shows, the three entrances along Liberty Street were used most extensively between 9 a.m. and 4 p.m. The southern entrance was used less often, except during the early morning hours and late in the afternoon; this is the major portal for employees arriving at and leaving work. Heavy use of the main entrance occurred between noon and 1 p.m. and around 10 a.m. The northeast entrance of the Post Office was used most extensively between 11 a.m. and 4 p.m., and its central entrance was used most heavily in the afternoon.

Users of the Post Office entrances were short-term visitors. Their primary purpose in coming to the building was to pick up mail, purchase stamps, or mail letters or packages. Many walked, but others arrived in cars and parked along Liberty Street or in the short-term lot.

*Where do community residents go*
*when they visit the Federal Building?*

Community residents contacted by telephone were asked if they had ever visited the Post Office or other governmental agencies in the Federal Building. Nine out of ten of those who had ever visited the building

FIGURE 6.1

Hourly Number of People Entering
and Leaving Building Each Hour of the Day

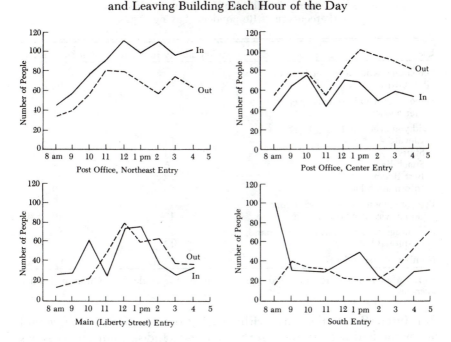

had been to the Post Office, while somewhat less than two-thirds (63 percent) had been to one of the other governmental agencies (Table 6.1). These proportions were similar among visitors contacted at the site. Clearly, the agency visited most often was the Post Office (Table 6.2). Other than the Post Office, the agencies most frequently visited were the Social Security Administration, the Internal Revenue Service, and the military recruiters. While the Social Security office has a consistent flow of visitors throughout the year, customer service at the Internal Revenue Service varies; at the time our observations were made, relatively few people were visiting IRS. We were told, however, that during the first quarter of each year, when people were preparing tax returns, customer service is higher than that provided by the Social Security Administration.

Visitors were also asked questions about their use of specific building facilities such as elevators, stairs, the snack bar, restrooms, and the information desk (Table 6.1). On-site visitors were much more likely than telephone respondents to say they had used each of these facilities.

TABLE 6.1

Use of the Federal Building
by Community Residents and On-Site Visitors
(Percentage of Respondents Saying "Yes")

| Uses | Community Residents | On-Site Visitors |
|------|---------------------|------------------|
| Percentage who have ever been to Post Office | 89 | 93 |
| Percentage who have ever been to other agencies | 63 | 71 |
| "Did you ever use the _____?" | | |
| Elevators | 21 | 43 |
| Stairs | 24 | 43 |
| Snack Bar | 5 | 17 |
| Rest Rooms | 12 | 45 |
| Information Desk | 20 | 20 |
| Percentage who have had difficulty finding way around building | 17 | 10 |
| Percentage who have wandered around or explored building | 13 | 23 |
| Number of respondents | 84 | 60[a] |

[a] Five respondents were not asked questions on the use of the Post Office and other agencies.

It is interesting to note that within each group, one person in five said he or she had been to the information desk. Indeed, our observations indicated that a number of people did seek assistance from a security guard stationed at the front desk. Based on our systematic counts, an average of 50 inquiries were made during the course of an eight-hour day, with most occurring during the noon hour. Some people clearly had problems in finding building facilities and specific agencies other than the Post Office. In fact, 17 percent of those contacted by telephone and 10 percent of the on-site visitors said they had difficulty in finding their way around the building. We asked the people who had used specific facilities if they had difficulty in finding them. Only one of the 28 people from both groups who had been to the information desk had problems in locating it. On the other hand, 6 of the 14 people who went to the snack bar on the second floor said they had difficulties in finding it.[2] Part of the orientation problem is related to inadequate signboards, an issue that will be addressed more thoroughly later in this chapter.

*How much do federal employees use the building?*

In order to develop an understanding of the extent to which people working in different agencies use the various building facilities and ser-

TABLE 6.2

Impressionistic Observations of Behavior within Selected Agencies—October/November 1979

| Behavioral Observations | Post Office | IRS | Military Recruiters | HCRS | Social Security | Weather Service |
|---|---|---|---|---|---|---|
| Customer service | high | low/average | average | none | high | none |
| Customer penetration | none | low | medium | none | high/medium | none |
| Staff interaction | average | low | average | high | average | average |
| Staff movement | high/average | average | average | average | high/average | high |

*Limited use was made of the snack bar and lounge area in the second-floor lobby. One-third of the employees had not been to the snack bar, while three-quarters had not used the lounge during the previous month.*

vices, we included in the employees' questionnaire several questions designed to ascertain the number of times during the month people had visited the conference room, the snack bar, the lounge outside the snack bar, the Post Office, and agencies other than their own. Among the locations considered, the conference room on the second floor was used least often (see Table 6.3). While virtually no one from the Weather Service or the Post Office used the conference room, others used it with some regularity. The most active users were the HCRS staff (3 times per month) and, to a lesser extent, the military recruiters (1.3 times per month).[3]

The snack bar and its adjacent lounge area received varied use by employees from different agencies. The typical HCRS employee went to the snack bar 6.3 times per month, and postal workers used it only about once a month.[4] The lounge area outside the snack bar was used most extensively by employees from the Social Security Administration (five times a month on average); others in the building used the lounge less than once a month. Observer counts confirmed the relatively limited use of these second-floor facilities. On average, about 170 daily visits were made to the snack bar, while the number of people who sat in the

TABLE 6.3

Employees' Monthly Use of Building Facilities/Services, by Agency[a] (Percentage Distribution)

| | | | | | Agency | | | |
|---|---|---|---|---|---|---|---|---|
| Monthly Visits to: | All | Post Office | IRS | Military Recruiters | HCRS | Social Security | Weather Service | Small Agencies |
| **Conference Room** | | | | | | | | |
| None | 63 | 98 | 57 | 46 | 11 | 61 | 100 | 81 |
| 1–2 times | 25 | 2 | 41 | 46 | 37 | 33 | — | 15 |
| 3–5 times | 10 | — | 2 | — | 43 | 6 | — | 4 |
| 5–10 times | 2 | — | — | 8 | 9 | — | — | — |
| More than 10 times | — | — | — | — | — | — | — | — |
| Total | 100 | 100 | 100 | 100 | 100 | 100 | 100 | 100 |
| Mean visits | 0.9 | — | 0.7 | 1.3 | 3.0 | 0.7 | — | 0.4 |
| **Snack Bar/Candy Shop** | | | | | | | | |
| None | 31 | 77 | 25 | 23 | 11 | 13 | 37 | 11 |
| 1–2 times | 22 | 13 | 18 | 23 | 15 | 39 | 37 | 26 |
| 3–5 times | 19 | 2 | 31 | 54 | 15 | 18 | 13 | 30 |
| 5–10 times | 16 | 6 | 10 | — | 35 | 18 | — | 26 |
| More than 10 times | 12 | 2 | 16 | — | 24 | 12 | 13 | 7 |
| Total | 100 | 100 | 100 | 100 | 100 | 100 | 100 | 100 |
| Mean visits | 3.7 | 1.2 | 4.2 | 2.5 | 6.3 | 4.1 | 2.6 | 4.4 |
| **Snack Bar/Lounge** | | | | | | | | |
| None | 72 | 93 | 84 | 85 | 64 | 30 | 87 | 66 |
| 1–2 times | 16 | 5 | 12 | 15 | 30 | 18 | 13 | 15 |
| 3–5 times | 4 | 2 | 2 | — | 2 | 15 | — | 4 |
| 5–10 times | 1 | — | — | — | 2 | 4 | — | 4 |
| More than 10 times | 7 | — | 2 | — | 2 | 33 | — | 11 |
| Total | 100 | 100 | 100 | 100 | 100 | 100 | 100 | 100 |
| Mean visits | 1.3 | 0.2 | 0.5 | 0.2 | 0.9 | 5.1 | 0.2 | 2.0 |

TABLE 6.3 (Continued)

| Monthly Visits to: | All | Post Office | IRS | Military Recruiters | HCRS | Social Security | Weather Service | Small Agencies |
|---|---|---|---|---|---|---|---|---|
| | | | | | Agency | | | |
| **Post Office** | | | | | | | | |
| None | 13 | 26 | 8 | 8 | 6 | 7 | 25 | 15 |
| 1–2 times | 26 | 31 | 33 | 8 | 21 | 36 | 25 | 11 |
| 3–5 times | 29 | 9 | 41 | 8 | 43 | 33 | 31 | 22 |
| 5–10 times | 17 | 17 | 12 | 38 | 21 | 9 | 6 | 22 |
| More than 10 times | 15 | 17 | 6 | 38 | 9 | 15 | 13 | 30 |
| Total | 100 | 100 | 100 | 100 | 100 | 100 | 100 | 100 |
| Mean visits | 4.6 | 4.1 | 3.8 | 7.9 | 4.7 | 4.3 | 3.6 | 6.3 |
| **Other Agencies** | | | | | | | | |
| None | 62 | 70 | 65 | 15 | 66 | 79 | 75 | 30 |
| 1–2 times | 23 | 26 | 33 | 15 | 26 | 12 | 19 | 19 |
| 3–5 times | 8 | 2 | – | 46 | 2 | 3 | 6 | 33 |
| 5–10 times | 4 | – | 2 | 9 | 4 | 3 | – | 11 |
| More than 10 times | 3 | 2 | – | 15 | 2 | 3 | – | 7 |
| Total | 100 | 100 | 100 | 100 | 100 | 100 | 100 | 100 |
| Mean visits | 1.3 | 0.7 | 0.6 | 4.5 | 1.0 | 0.9 | 0.5 | 3.3 |
| Number of respondents | 239 | 54 | 49 | 13 | 47 | 33 | 16 | 27 |

[a] The question, "During the past month how many times have you ____?", was asked about six facilities/services. Responses to the item "asked assistance from the security guard" are not presented; on average, responses to this item were less than once a month; two-thirds of the employees did not ask for assistance at all.

lounge area was considerably smaller.[5] In part, the proximity of down-town restaurants and grocery stores and the snack bar's limited hours of operation contributed to its low use; at the time the data were being collected, it opened at 8:30 a.m. and closed anytime between 1:30 p.m. and 3:30 p.m. Its relatively low use can also be explained by the fact that the public did not readily know of its existence; directions to the second-floor facility were poor. The infrequent use of the lounge area was probably related to the limited quantity and low quality of its fur-nishings; indeed, the lounge furniture did not resemble that specified as part of the original plan by the interior designer.

For the most part, Federal Building employees have limited contact with agencies other than their own or the Post Office. Nearly everyone had purchased stamps or mailed letters or packages in the building, but less than four in ten had been to other agencies.

In order to measure the full extent to which the building was being used by its employees, we created a composite index for building use (an index score for each employee was built from responses to the ques-tions dealing with the frequency of use of facilities and services in the building). This procedure showed that the military recruiters, HCRS employees, and those in the small agencies were the most active build-ing users. Least likely to use building facilities and services were the postal workers and Weather Service personnel (Table 6.4).[6]

*What factors account for how extensively
the federal employees use the building?*

It was hypothesized that employees who were relatively new to their jobs would use the building facilities less extensively than those who had been working in the building since it opened. Indeed, an examina-tion of building-use data by length of employment at the building showed that a modest relationship did exist. However, that relation-ship was not in the expected direction. People who had worked in the building for relatively short periods of time (less than six months) were more likely than others to use the facility.

In order to see if this relationship was maintained after taking into account job classification and agency, a multiple classification analysis predicting to building use was performed. Clearly, the most important factor was the agency in which the individuals worked (Table 6.5). The adjusted mean scores show that, after taking into account both type of job and length of tenure in the building, HCRS employees used the building most often, and postal workers used the building consider-ably less than average. Within agencies, differences in job classification helped to explain building use. For example, managerial personnel in

TABLE 6.4

Employees' Use of the Building, by Agency
(Percentage Distribution)

| Building Use Index[a] | All | Agency | | | | | | |
|---|---|---|---|---|---|---|---|---|
| | | Post Office | IRS | Military Recruiters | HCRS | Social Security | Weather Service | Small Agencies |
| High (5) | 15 | 2 | 4 | 39 | 24 | 24 | 6 | 26 |
| (4) | 20 | 7 | 16 | 31 | 36 | 28 | – | 22 |
| (3) | 23 | 13 | 39 | 15 | 19 | 18 | 25 | 30 |
| (2) | 22 | 22 | 27 | 15 | 17 | 15 | 38 | 22 |
| Low (1) | 20 | 56 | 14 | – | 4 | 15 | 31 | – |
| Total | 100 | 100 | 100 | 100 | 100 | 100 | 100 | 100 |
| Number of respondents | 239 | 54 | 49 | 13 | 47 | 33 | 16 | 27 |
| Mean score | 2.9 | 1.8 | 2.7 | 3.9 | 3.6 | 3.3 | 2.1 | 3.5 |

[a] For each individual, an index score was created using responses to questions dealing with the frequency of use of facilities/services in the building. The question, "During the past month, how many times have you visited ____?", was asked about six places: the conference room, the snack bar/candy shop, the lounge, the Post Office, another agency and the security guard's desk. Responses ranged from none to more often than ten times during the past month. The higher the score, the greater the likelihood that the individual uses the building.

TABLE 6.5

Federal Building Use Predicted by Selected Employee Characteristics
(Multiple Classification Analysis; N = 231)

| Predictors | Eta Coefficient | Beta Coefficient | Adjusted Use Scores |
|---|---|---|---|
| Agency | .54 | .54[a] | Postal workers (1.80); HCRS (3.67) |
| Job Classification | .42 | .18 | Clerical-secretarial (2.74); military recruiters (3.82); managers-supervisors (3.12) |
| Time in Building | .00 | .17 | Little difference |
| Percentage of variance explained (adjusted multiple $R^2$) | | 30.0 | |
| Overall Use Score = 2.86[b] | | | |

[a] The predictors are listed in order of importance.

[b] The higher the use score, the greater the likelihood that the individual uses the building.

HCRS were likely to use the building more extensively than clericals or secretaries. Finally, length of time in the building did not help to explain building use. Employees who had worked in the building for only short periods were just as likely to use the facilities as people who had worked there for more than six months.

## Evaluating the Building

In the questionnaires administered to community residents and to on-site visitors, respondents were asked about aspects of the building they particularly liked or disliked. Table 6.6 shows that more than half of the telephone respondents and two-thirds of the on-site visitors said there was indeed something about the building they particularly liked.[7] Community residents most often mentioned liking the plaza, including its landscaping and paving, the building's setback from the street, and the stairs. Many people also liked the architectural design and the attractive windows and skylights. People interviewed at the building site most often mentioned attributes related to the plaza, the overall building design, its spacious interior, and the location.

About one-third of each of the three groups also indicated there was something about the building they disliked. Telephone respondents most often said they disliked the overall design; they also thought the location was poor and they disliked exterior features such as the drab color or the materials used. Among people contacted at the building it-

TABLE 6.6

Building Attributes Liked and Disliked
by Community Residents and On-Site Visitors
(Percentage Distribution)

| Building Attributes | Community Residents | On-Site Visitors |
|---|---|---|
| "Is there anything about the building you especially like?" | | |
| Percentage saying "yes" | 55 | 67 |
| Number of respondents | 103 | 60 |
| Building Attributes Liked | | |
| Plaza: planters, trees, landscaping, paving pattern, setback, entrance, stairs | 33 | 16 |
| Overall design: modern, attractive, good shape, imaginative, new | 20 | 18 |
| Glass and windows: tiered, skylights | 19 | 11 |
| Location: convenient, fits in downtown well | 13 | 16 |
| Exterior: brickwork, color | 5 | 4 |
| Interior: spacious, clean | 3 | 18 |
| Service/personnel: friendly people; good hours | 1 | 7 |
| Other: well built, parking convenient | 5 | 10 |
| Mention of something disliked | 1 | — |
| Total | 100 | 100 |
| Number of mentions | 75 | 45 |
| Number of respondents | 57 | 40 |
| "Is there anything about the building you especially dislike?" | | |
| Percentage saying "yes" | 38 | 33 |
| Number of respondents | 103 | 60 |
| Building Attributes Disliked | | |
| Overall design: too modern, looks like factory, plain | 37 | 5 |
| Exterior: too much brick, color dull, bare walls | 14 | 5 |
| Glass and windows: too many facing light, not enough | 12 | 9 |
| Parking: inconvenient, not enough spaces | 12 | 19 |
| Other: poor location, poor entrance, interior | 18 | 62 |
| Service/personnel: poor, security bad | 5 | — |
| Mention of something liked | 2 | — |
| Total | 100 | 100 |
| Number of mentions | 45 | 22 |
| Number of respondents | 40 | 20 |

*When asked what they liked best about the Federal Building, community residents most often mentioned the plaza, including its landscaping, paving, and general decor. One out of three persons had used it at one time or another.*

self, the most frequent complaints were about its location and the inconvenience of parking. Both the positive and negative features mentioned by the two groups are shown in Table 6.6.

*Does the public think the building is attractive?*

The above findings suggest that Ann Arbor residents have mixed responses to the building. However, the ratio of positive to negative responses indicates that, on balance, the public viewed the building favorably. Indeed, most of the persons questioned did say they thought the building is attractive (Table 6.7). Among those contacted by telephone, two-thirds said the interior was attractive, three-quarters felt the exterior of the building was attractive, and close to 90 percent thought the plaza in front of the building was attractive.[8] Among people contacted at the site, eight in ten said the interiors were attractive. An identical proportion rated the exterior in that manner, and nearly everyone thought the plaza was attractive.

In order to see if any particular group was more or less likely to rate the building favorably, average ratings were calculated for different subgroups of telephone respondents. Although none of the differences in the mean scores were statistically significant, Table 6.8 shows that respondents who were not U-M students, who had lived in Ann Arbor for more than 12 years, and who did not work in the central area gave higher marks to the building and the plaza. At the same time, familiarity with the building in terms of the number of visits tended to be associated with lower ratings.

*How do occupants rate the building as a place to work?*

The federal employees were asked to evaluate the building from the point of view of its attractiveness, its architectural quality, its upkeep, and, more generally, as a place to work. In response to the question, "Overall, how would you rate the building as a place to work?" 10 percent rated it as excellent, 46 percent said very good, 28 percent said it was fair, and the remainder (16 percent) said it was a poor place to work. Highest ratings were given by the military recruiters and employees in the small agencies. Among all building occupants, those in HCRS, the Weather Service, and the Social Security Administration were most likely to give the building low marks (see Table 6.9). As we shall see, people's responses to the question were strongly associated with their feelings about the overall architectural quality of the building and about their own work environments.

The federal employees tended to give more favorable ratings to specific attributes of the building than to the building itself. Table 6.10

TABLE 6.7

Ratings of Federal Building Attractiveness
by Community Residents and On-Site Visitors
(Percentage Distribution)

| Ratings | Community Residents | On-Site Visitors |
|---|---|---|
| Interior Appearance[a] | | |
| Very attractive | 16 | 25 |
| Fairly attractive | 48 | 55 |
| Not very attractive | 28 | 18 |
| Not at all attractive | 8 | 2 |
| Total | 100 | 100 |
| Number of respondents | 77 | 60 |
| Mean rating[b] | 2.7 | 3.0 |
| Exterior Appearance[c] | | |
| Very attractive | 31 | 36 |
| Fairly attractive | 44 | 43 |
| Not very attractive | 19 | 17 |
| Not at all attractive | 6 | 4 |
| Total | 100 | 100 |
| Number of respondents | 98 | 58 |
| Mean rating | 3.0 | 3.1 |
| Plaza[d] | | |
| Very attractive | 44 | 41 |
| Fairly attractive | 43 | 54 |
| Not very attractive | 12 | 5 |
| Not at all attractive | 1 | — |
| Total | 100 | 100 |
| Number of respondents | 97 | 60 |
| Mean rating | 3.3 | 3.4 |

[a] The question was: "What do you think of the appearance of the inside of the building? Would you say it's very attractive, fairly attractive, not very attractive, or not at all attractive?"

[b] Mean ratings are based on scores of 4 for "very attractive," 3 for "fairly attractive," 2 for "not very attractive," and 1 for "not at all attractive."

[c] The question was: "And what do you think about the overall appearance of the outside of the building? Is it very attractive, fairly attractive, not very attractive, or not at all attractive?"

[d] The question was: "What about the plaza or open area in front of the building? Would you say it's very attractive, fairly attractive, not very attractive, or not at all attractive?"

shows that the building's upkeep and the convenience of its restrooms were generally evaluated positively, whereas low ratings were given to the attractiveness of signs and to several dimensions of architectural quality. The poor ratings of the signs were not surprising. At the time the questionnaires were being administered, permanent signs had not

TABLE 6.8

Ratings of Federal Building Attractiveness,
by Characteristics of Community Residents
(Mean Ratings)[a]

| Characteristics | Ratings of: | | |
| --- | --- | --- | --- |
| | Interior Appearance | Exterior Appearance | Plaza |
| Total Sample | 2.7(77)[b] | 3.0(98) | 3.3(97) |
| Residential Location | | | |
| Central Ann Arbor | 2.7( 7) | 2.9(10) | 3.2(10) |
| Elsewhere | 2.7(70) | 3.0(88) | 3.3(87) |
| Length of Residence in Ann Arbor | | | |
| 2 years or less | 2.5(15) | 2.9(21) | 3.0(22) |
| 3–12 years | 2.7(28) | 3.2(39) | 3.3(37) |
| Longer than 12 years | 2.8(32) | 3.1(36) | 3.5(36) |
| Student Status | | | |
| U-M student | 2.6(14) | 2.7(16) | 3.1(18) |
| Non-student | 2.8(61) | 3.1(79) | 3.4(76) |
| Place of Employment | | | |
| Downtown Ann Arbor | 2.0( 7) | 2.8( 9) | 3.1( 9) |
| U-M campus | 2.5(17) | 2.6(19) | 3.1(18) |
| Elsewhere | 3.0(18) | 3.2(26) | 3.4(26) |
| Number of Downtown Visits Last Month | | | |
| None | 3.0( 3) | 3.0( 6) | 3.2( 5) |
| 1–4 | 2.7(22) | 3.0(30) | 3.0(28) |
| 5–10 | 2.9(21) | 3.2(27) | 3.5(27) |
| 11–20 | 2.7(14) | 2.9(17) | 3.1(18) |
| More often | 2.3(15) | 2.8(15) | 3.1(16) |
| Visited Building Last Month | | | |
| Yes | 2.8(35) | 2.9(39) | 3.4(38) |
| No | 2.6(42) | 2.9(41) | 3.3(44) |
| Sex | | | |
| Male | 2.7(35) | 2.9(45) | 3.2(47) |
| Female | 2.7(42) | 3.1(52) | 3.4(44) |

[a] Mean ratings are based on scores of 4 for "very attractive," 3 for "fairly attractive," 2 for "not very attractive," and 1 for "not at all attractive."

[b] Numbers in parentheses report the number of respondents in each group.

been installed and a variety of temporary placards were being used throughout the building. The extent to which different agencies rated these various attributes positively or negatively is shown in Table 6.10.

Several employee characteristics were examined to see how well they explained the different evaluations of specific building attributes. Using bivariate analyses, we found that two factors were significant

TABLE 6.9

Rating of Federal Building as a Place to Work, by Agency
(Percentage Distribution)

| "Overall, how would you rate the building as a place to work?" | All | Agency | | | | | | |
| --- | --- | --- | --- | --- | --- | --- | --- | --- |
| | | Post Office | IRS | Military Recruiters | HCRS | Social Security | Weather Service | Small Agencies[a] |
| Excellent | 10 | 15 | 4 | 39 | 4 | 12 | – | 7 |
| Pretty good | 46 | 46 | 49 | 46 | 26 | 40 | 50 | 78 |
| Fair | 28 | 30 | 30 | 15 | 38 | 27 | 25 | 15 |
| Poor | 16 | 9 | 17 | – | 32 | 21 | 25 | – |
| Total | 100 | 100 | 100 | 100 | 100 | 100 | 100 | 100 |
| Number of respondents | 237 | 54 | 47 | 13 | 47 | 33 | 16 | 27 |
| Mean rating[b] | 2.5 | 2.7 | 2.4 | 3.2 | 2.0 | 2.4 | 2.2 | 2.9 |

[a] Includes these agencies: Department of Defense-Army Recruiting Area Commander; Defense Logistics Agency; Defense Investigative Service; and Army Surgeon General; Soil Conservation Service; District Court-Probation Department; Department of Labor, Wage and Hourly Division; the Federal Bureau of Investigation; and the security guard.

[b] Mean ratings are based on scores of 4 for "excellent," 3 for "pretty good," 2 for "fair," and 1 for "poor."

TABLE 6.10

Evaluation of Building Attributes, by Agency
(Mean Scores)[a]

| Building Attributes | All | Agency | | | | | | |
| --- | --- | --- | --- | --- | --- | --- | --- | --- |
| | | Post Office | IRS | Military Recruiters | HCRS | Social Security | Weather Service | Small Agencies |
| Well/poorly kept up exterior | 5.4 | 5.8 | 5.0 | 6.4 | 4.9 | 5.0 | 5.5 | 6.0 |
| Conveniently/inconveniently located toilets | 5.3 | 5.5 | 3.9 | 6.5 | 5.8 | 5.5 | 5.7 | 5.9 |
| Well/poorly kept up interior | 5.0 | 5.5 | 4.3 | 6.2 | 4.5 | 5.4 | 4.6 | 5.5 |
| Easy/difficult to find way | 4.6 | 4.8 | 4.6 | 5.2 | 3.9 | 4.3 | 5.4 | 5.0 |
| Pleasant/unpleasant | 4.6 | 5.0 | 4.3 | 5.6 | 3.7 | 4.9 | 4.3 | 5.1 |
| Attractive/unattractive | 4.5 | 5.1 | 4.1 | 5.6 | 3.6 | 4.5 | 5.1 | 4.7 |
| Excellent/poor security | 4.1 | 4.2 | 4.2 | 5.5 | 3.6 | 4.6 | 3.8 | 3.4 |
| Good/poor architectural quality | 3.8 | 4.5 | 3.7 | 4.8 | 2.9 | 3.4 | 2.8 | 4.9 |
| Good/poor design | 3.6 | 4.5 | 3.5 | 4.5 | 2.5 | 3.4 | 2.6 | 4.4 |
| Attractive/unattractive signs | 3.1 | 4.5 | 3.0 | 2.5 | 2.8 | 2.2 | 3.6 | 2.5 |
| Stimulating/unstimulating spaces | 3.0 | 3.4 | 2.5 | 4.7 | 2.1 | 3.4 | 3.5 | 3.2 |
| Number of respondents | 238 | 53 | 49 | 13 | 47 | 33 | 16 | 27 |

[a] Building attributes were rated on a scale from 1 to 7; the higher the number the more favorable the rating. Scores above 4 are considered positive, and scores below 4 are considered negative.

*Differences were found in the quality of signs used on the outside of the building and those used indoors. Interior signs were rated poorly by most federal employees. Feelings about the attractiveness of signs influenced people's assessments of the building's architectural quality.*

predictors — the employee's job classification and the length of time the employee had worked in the building. Generally, the military recruiters and postal workers gave the highest ratings, while managerial/supervisory and professional/technical personnel gave the lowest ratings. Employees who had worked in the building for less than one year tended to rate specific attributes more positively than those who had been in the building for a year or longer.

Several of the building attributes we asked about were intended to represent various dimensions of a single concept. For example, views on architectural quality were to be tapped through ratings of the building's design, its attractiveness, the extent to which its spaces were stimulating, and a single item on architectural quality. In a correlational analysis, these four dimensions, along with ratings of pleasantness, were shown to be highly interrelated (see Appendix Table A.1). Accordingly, an index of architectural quality was created to try to capture the multidimensional nature of the concept. Similarly, an index was created to represent people's views about the upkeep of the building. The average ratings given to these two dimensions by employees from different agencies are shown in Table 6.11.[9]

Employees clearly had mixed feelings about the architecture of the building. These varied sentiments were also revealed in the comments volunteered by people completing their questionnaires. One person wrote:

> I think that architecturally this building is exciting and different and I'm proud to work in it.

Another employee thought the building was "worth its weight in gold." Other persons, however, were not so kind in their added remarks:

> If this building won an award for design excellence, I would hate to work in other federal buildings.

> I resent the fact that the designers received an award without the benefit of employee input.

*How do employees' feelings about specific building attributes influence their overall ratings of the building as a place to work?*

We used a regression analysis to answer this question. As explanatory or predictor variables, we used employees' ratings of the attractiveness of signs, the convenience of toilets, the ease of finding one's way, and security, along with the indexes of architectural quality and maintenance. The evaluations of these six factors accounted for 47 percent of the variability in the way the federal employees rated the building as a

TABLE 6.11

Ratings of Building's Architectural Quality and Upkeep, by Agency
(Mean Score)[a]

| Agency | Ratings of: | | Number of Respondents |
|---|---|---|---|
| | Architectural Quality[b] | Upkeep[c] | |
| Military Recruiters | 26.3 | 11.2 | 13 |
| Small Agencies | 22.4 | 10.5 | 27 |
| Post Office | 21.9 | 10.3 | 53 |
| Social Security | 19.2 | 9.3 | 33 |
| Weather Service | 18.2 | 9.1 | 16 |
| IRS | 18.1 | 8.3 | 49 |
| HCRS | 15.1 | 8.4 | 47 |
| All | 19.4 | 9.4 | 238 |

[a] Mean scores for architectural quality ranged from 7 to 35; for upkeep, they ranged from 2 to 14. The higher the number, the higher the quality rating.

[b] Architectural quality is an index created from individual responses to five items dealing with the building's attractiveness, design, pleasantness, architectural quality, and the extent to which spaces are stimulating. For a review of the inter-item correlations, see Appendix Table A.1.

[c] "Upkeep" is an index created from individual responses to two items dealing with interior and exterior maintenance. The inter-item correlations are reported in Appendix Table A.1.

place to work. By far the most important factor in predicting people's overall rating was their assessment of the architectural quality of the building. Feelings about architectural quality, the only significant predictor among those considered, accounted for 42 percent of the total variance.[10]

*What factors account for differences in people's feelings about the architectural quality of the Federal Building?*

There are undoubtedly numerous factors in addition to architectural quality that contribute to people's assessments of a building as a place to work. In fact, we found that, relative to their views on architectural quality, people's feelings about their agency and their specific work environments were more strongly associated with their overall ratings. This would suggest that there are both functional and aesthetic dimensions to the responses of people who are asked to give a general rating to the building in which they work. Our purpose here is not to examine the exact nature of each of these dimensions; that would require a different set of measures than those available as part of this evaluation. Suffice it to say that, as our analysis shows, views on architectural quality are important in understanding the way buildings are judged by the people who occupy them.

We were interested in knowing not only what the occupants thought about the Federal Building from an architectural point of view, but also how those thoughts might differ among various occupants. We noted earlier that, in addition to differences among the employees of different agencies in their ratings of architectural quality, ratings differed for people who had various jobs and who occupied the building for varying lengths of time. As Table 6.12 shows, these three factors taken together accounted for nearly a quarter of the variation in the evaluative scores on architectural quality. Even after taking into consideration the types of jobs and length of building occupancy, agency differences were still salient in predicting employees' feelings about the architecture. The military recruiters, the personnel in the smaller agencies, and the postal workers — irrespective of job level or duration of employment in the building — gave the highest ratings, and HCRS personnel gave the lowest.

As a way of determining whether particular building attributes related to architectural quality contributed to the ratings, employees' feelings about the building's signs and upkeep were considered in a second multiple classification analysis. As shown in the second part of Table 6.12, this procedure increased the proportion of explained variance to 35.7 percent. After taking into account who the employees were, their ratings of the building upkeep was second only to their agency affiliation in explaining their feelings about its architecture.

In order to further understand why the employees' agencies were so important to their views on architectural quality, we decided to investigate that quality vis-à-vis individuals' feelings about their agencies. Two ratings of the agency workspaces themselves are discussed in a subsequent chapter dealing with the evaluation of the work environment. The ratings of general ambience seemed most appropriate for our explorations of architectural quality.[11] When agency ambience was considered along with the other predictors, the proportion of variance accounted for in a multiple classification analysis increased to 45 percent. As the Beta coefficients indicate, agency ambience became the most important predictor. Second in importance was the agency of the respondent. After all other predictors were taken into account, the building's architecture was shown to be least satisfying to HCRS personnel, irrespective of how they rated the general ambience of their agency. These data strongly suggest that the way people view their work environments can significantly color their feelings about the building as a whole, including the quality of its architecture. This point will be discussed later in more detail.

TABLE 6.12

Evaluation of Architectural Quality Predicted by
Ratings of Building Attributes and Agency Ambience
(Multiple Classification Analysis—238 Employees)

| Predictor | Eta Coefficient | Beta Coefficient[a] | | |
| --- | --- | --- | --- | --- |
| | | Employee Characteristics | Employee Characteristics and Building Attribute Ratings | Employee Characteristics, Building Attribute, and Agency Ambience Ratings |
| Employee Characteristics | | | | |
|   Agency | .45 | .36(1) | .34(1) | .25(2) |
|   Time in building | .26 | .27(2) | .17(5) | .18(6) |
|   Job classification | .36 | .20(3) | .26(3) | .25(3) |
| Attribute Ratings | | | | |
|   Signs | .29 | | .21(4) | .20(5) |
|   Upkeep | .42 | | .29(2) | .22(4) |
| Agency Ambience Rating | .54 | | | .35(1) |
| Percentage of variance explained (adjusted multiple $R^2$) | | 24.1 | 35.7 | 44.7 |

[a] Numbers in parentheses indicate ranking of importance.

## Notes

1. See Chapter 2 for a discussion of the methods used in conducting the systematic observations at the building entrances.

2. Ten of 36, 5 of 46, and 4 of 43, respectively, reported difficulty in locating the restrooms, the stairways, and elevators.

3. The Weather Service and the Social Security Administration had space set aside within their agencies for group meetings.

4. It should be noted that a lounge area including food dispensing machines was available within the Post Office space.

5. Admittedly, the figure on snack bar use is a rough and liberal estimate based on incomplete sample counts and an assumption about equal use for each hour during the period it was open. Estimates on lounge use were even rougher and therefore are not reported. The military recruiters used the building most extensively and the postal carriers used it marginal operation.

6. Building use was also examined for employees with different job classifications. The military recruiters used the building most extensively and the postal carriers used it least often; managerial, professional/technical and clerical/secretarial personnel were all comparable in their use of the building facilities.

7. The question was not asked of community residents who did not know where the building was located.

8. The question about the interior was only asked of telephone respondents who said they had been to the building. Questions about the exterior and the plaza were not asked of respondents who did not know where the building was located.

9. An examination of the index scores for people with different job classifications and different lengths of tenure in the building revealed relationships similar to those reported earlier. The lowest ratings were given by managerial, supervisory, and professional/technical personnel, and the highest ratings were given by the military recruiters. Long-term occupants of the building were most likely to give poor ratings to its architectural quality and upkeep.

10. Although the remaining predictors contributed little to our understanding of the way people rated the building as a place to work, ease of finding one's way was the second most important predictor, and people's feelings about the convenience of toilets was least important.

11. For a detailed discussion of the measure of agency ambience dealing with the way the agency looks and the extent to which it is pleasant, see Chapter 7.

# 7

# Evaluating the Work Environment

## Overview

Both in the popular press and among the research community, considerable attention has been given in recent years to the work environment as it relates to the quality of working life, productivity issues, corporate image, and worker benefits. We were not surprised, then, to find that much of the discussion with GSA representatives, with the architects, and with the heads of the various federal agencies focused on the work environment and people's responses to it. During these discussions, it became apparent that the concept of "work environment" is multidimensional, having both psychological and organizational components, as well as physical components. It was apparent from our preliminary review of the data that these components were related in the minds of the people working in the Federal Building.

For the purposes of this evaluation, we have nonetheless defined the work environment in physical terms and, within the context of the Ann Arbor Federal Building, we have viewed it as operating at two levels: at the organizational level and at the level of the individual work stations. Fourteen different organizations — separate federal agencies — are housed in the Ann Arbor Federal Building. These agencies contain a total of 265 work stations. The work station is typically represented by either a single office, a desk and its immediate surroundings, or a particular space used by an individual in the performance of job-related tasks. In this chapter we first describe the functions of each agency and its spatial arrangements and general decor. Quantitative data on specific characteristics of the work stations are then presented. Finally, we examine people's assessments of their own agency's work environment and their individual work stations.

We found that, in general, employees in the Ann Arbor Federal Building were dissatisfied with both the general ambience and functional arrangement of their agencies. Poor air quality and noise from other agencies were associated with negative feelings about agency ambience, while distractions caused by the movements of people and furniture were important determinants of the way employees judged the agency's functional organization. People's feelings about their agencies were to a large extent influenced by their views about their immediate work environment.

Federal Building workers were of mixed minds in their assessments of the work places they occupied. While many expressed some level of satisfaction with their own work stations, a third indicated they were dissatisfied. Indeed, most felt their immediate work environment was worse than what they had experienced before moving to the Federal Building. The federal employees also gave poor marks to specific work station attributes. For the most part, they were unhappy with their ability to carry on conversations in private, with views to the outside, and with the number and location of electrical outlets. Relatively favorable ratings were given to co-worker access and the lighting situation. The ratings of work stations were particularly low in comparison to national data. For the most part, poor ratings were associated with limited workspace and with the type of office a person occupied. People who shared an open office and who occupied less than 60 square feet of space were least satisfied with their work station.

## Describing the Work Environment

As we indicated earlier, the 14 separate federal agencies in the building employed approximately 270 people. The six largest agencies include the Post Office, the Internal Revenue Service, the military recruiters, the Heritage Conservation and Recreation Service, the Social Security Administration, and the National Weather Service. The eight smaller agencies each had from two to seven employees.[1]

*Post Office.* Most employees of the U.S. Postal Service are housed in a large, open area on the ground floor of the building. In addition to this main workspace, there is a small supervisor's office, a lounge containing food dispensing machines, and two locker rooms with toilets. Small, free-standing mail sorting carrels are provided for each postal carrier in the central part of the Post Office area, with specialized sorting stations located at the north end of the area adjacent to the public mail boxes (see Figure 7.1). There are five customer service stations at the northwest corner of the area, facing the building's central lobby. The entire space occupies approximately 10,000 square feet and is connected to a loading dock with two sets of double doors on the south side of the building.

The Post Office is unique among the tenants of the building in that it typifies an industrial rather than an office environment in both function and appearance. The surfaces of the space are unfinished concrete, painted masonry, and exposed structural steel. No windows or skylights are present in the agency's work areas.

Work in the Post Office begins at 6 a.m. with mail sorting and bulk delivery and lasts until 5 p.m. when the public service counter is closed. The most active work periods occur between 6 and 9 a.m. when the carriers are sorting their route mail and postal clerks are sorting box mail and servicing the customer counter.

*Internal Revenue Service (IRS).* The 9,100 square feet of space occupied by the IRS is on the opposite side of the main floor lobby from the Post Office; this space is typical of an open-office environment. Because the space had been designed to eventually house a federal district court, the ceiling height is 12 feet, approximately three feet higher than other spaces in the building. The primary workspace arrangement consists of five-foot high moveable partitions and individual worker carrels (see Figure 7.2). A public waiting area is adjacent to the northeast entry from the lobby. Public access is restricted beyond this point for purposes of security and privacy. Tax auditors occupy the north half of the area and conduct private interviews in the individual carrels. In the south half, space is devoted to single-occupancy work stations for administrators and multiple-occupancy stations used by field agents on a rotating basis. The only window in the agency is along the western wall, visible from only half a dozen work stations. A skylight runs the length of the north wall, and there is an open lightwell in the center of the agency's area, opening to the Heritage Conservation and Recreation Service on the second-floor above.

The entire area is carpeted and treated with acoustic materials on the walls, partitions, and ceilings. Except for times of seasonally high workloads, agency personnel are active from 8 a.m. until 4:30 p.m.

*Military Recruiters.* Situated in four separate offices adjacent to the south lobby, the Army, Navy, Air Force, and Marine recruiters are housed in open-office arrangements. These agencies together occupy

### FIGURE 7.1
Post Office

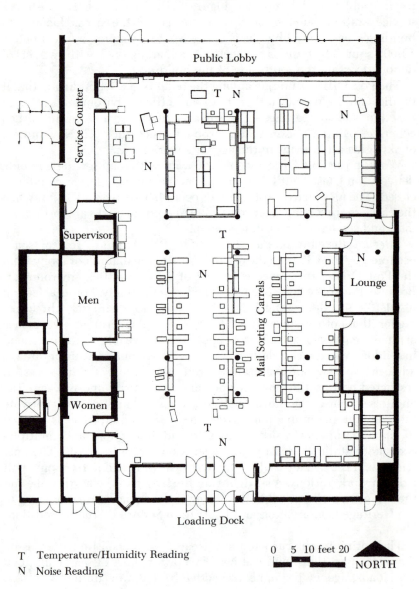

T   Temperature/Humidity Reading
N   Noise Reading

## FIGURE 7.2

### IRS and Military Recruiters

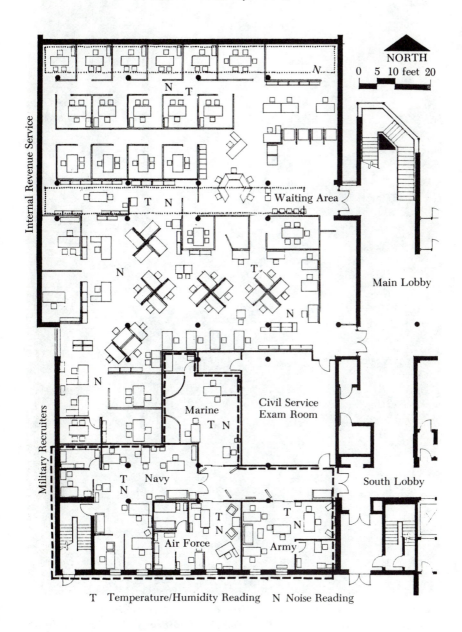

T  Temperature/Humidity Reading  N  Noise Reading

*Lightwells are characteristic of the spaces occupied by several agencies, including the Internal Revenue Service (above), where the lightwell is open to another agency on the floor above.*

*In the Social Security Administration Offices (above), a lightwell runs from the customer waiting area to a partitioned area toward the rear of the agency.*

approximately 1,000 square feet of space, all with 12-foot high ceilings. Although each office contains from three to six recruiters, there are no partitions separating the individual work stations (see Figure 7.2). The offices occupied by the Navy, Air Force, and Army each have two small, circular windows facing south, and the area assigned to Marine recruitment has the largest amount of space per person. Like the IRS, these agencies have acoustic finishes on the floors, walls, and ceilings.

The four offices are accessed by the public from an interior lobby adjacent to the south entry. The recruiters are open to the public from 8 a.m. to 6 p.m., and many of them conduct business outside the building thoughout the work day and on weekends.

*Heritage Conservation and Recreation Service (HCRS).* Occupying the largest area in the building (10,500 square feet), this agency is located on the western half of the second floor, adjacent to the large stairway lobby. Its spatial arrangement is similar to that of the IRS; work stations are separated by five-foot high moveable partitions, with no conventional or private offices (see Figure 7.3). It is also finished with acoustic wall and ceiling finishes and is completely carpeted. The ceiling height here and in other offices on the second, third, and fourth floors is nine feet.

This agency has a full-height window along the entire length of the north wall and has a lightwell opening into the IRS space below. It also has an open lightwell running the length of the agency and connecting to the Weather Service space above.

For the most part, this agency employs managerial and research personnel who perform administrative tasks at their work stations. Many of its employees are also involved in projects outside the building. Support staff, including secretaries and draftsmen, also have their own work stations. The agency is open to the public from 8 a.m. to 5 p.m.

*Social Security Administration.* With offices on the second floor adjacent to the public elevator lobby, most Social Security employees are housed in a pool-office arrangement. Although three supervisors occupy private offices on the south side of the agency, the remaining personnel perform their work at desks with no visual or acoustic screening (see Figure 7.4). The agency occupies approximately 6,700 square feet of space and has a full-height window along the entire north wall. A lightwell runs from the customer waiting area to a partitioned area devoted to staff meetings and coffee breaks. Acoustic treatment is provided on the walls and ceiling and the floor is carpeted.

Employees of the Social Security Administration have extensive contact with the public, and most deal directly with clients from their

## FIGURE 7.3

### Heritage Conservation and Recreation Service

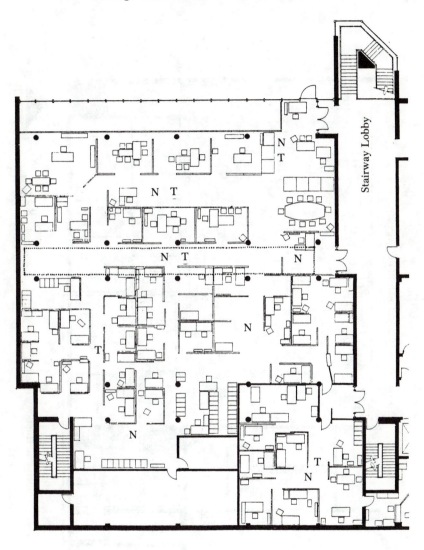

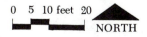

T    Temperature/Humidity Reading
N    Noise Reading

0   5  10 feet  20

NORTH

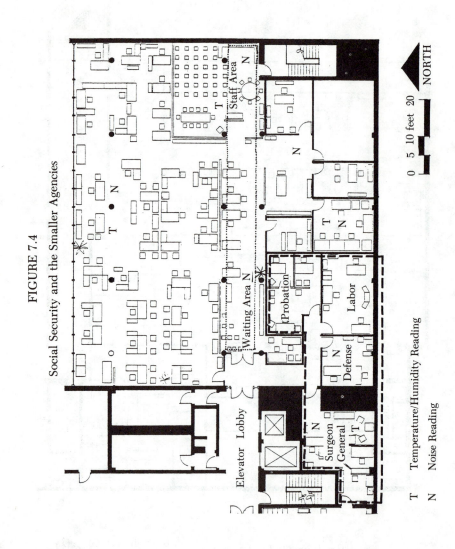

FIGURE 7.4

Social Security and the Smaller Agencies

T    Temperature/Humidity Reading

N    Noise Reading

desks. The agency is open for business between 9 a.m. and 4:30 p.m., although personnel work in the building from 7 a.m. to 6 p.m.

*National Weather Service.* Occupying 3,750 square feet on the third floor, this agency is unique in the work it performs. Except for one secretary/receptionist, the staff is comprised of meteorologists and weather technicians who operate a wide array of technical equipment and monitor a large number of communication devices. The equipment is staffed 24 hours a day, seven days a week. The area has a full-height window and an open lightwell to HCRS below. There is one private office and a conference room on the east and another enclosed room on the west for equipment repair personnel and for storage. The remainder of the area is subdivided with moveable partitions and forecasting machinery (see Figure 7.5).

*Small Agencies.* Occupying approximately 5,000 square feet in offices scattered throughout the building, seven of the eight small agencies occupy conventional offices which are often shared by two or three employees. Few have views to the outside, but all are carpeted and have acoustic wall and ceiling treatments. Only the fourth-floor offices of the Soil Conservation Service are open and have windows facing both north and south (see Figure 7.6). The Federal Bureau of Investigation, like the small agencies on the second floor, has no windows or skylights. It occupies a highly secure area on the third floor and public access is controlled by a receptionist at the entry.

### The Work Stations

In each agency, staff members perform specified tasks at their desks or other types of work stations. In total, there are 265 work stations in the building, although, at the time the questionnaires were being administered, not all of them were occupied. Similarly, some were shared by two or more people, while others were occupied by individuals who did not respond to the questionnaire. Environmental data covering the work stations were gathered ten weeks after the questionnaires were administered and have been organized according to six general categories: (1) type of work arrangement; (2) size and density of the work station; (3) work station furnishings and equipment; (4) accessibility of the work station to selected attributes of the agency; (5) lighting conditions; and (6) ambient environmental conditions such as temperature, humidity, and noise levels.

*Type of work arrangement.* The kinds of work arrangements or work stations in the building were determined for the evaluation in three ways. One approach involved a series of drawings presented to the employees as part of the questionnaire. The accompanying

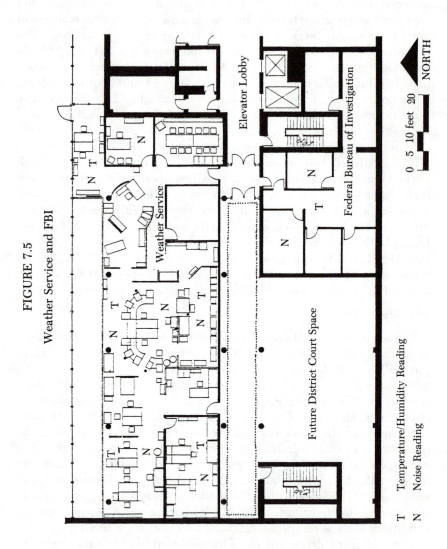

FIGURE 7.5

Weather Service and FBI

T   Temperature/Humidity Reading

N   Noise Reading

FIGURE 7.6

The Smaller Agencies

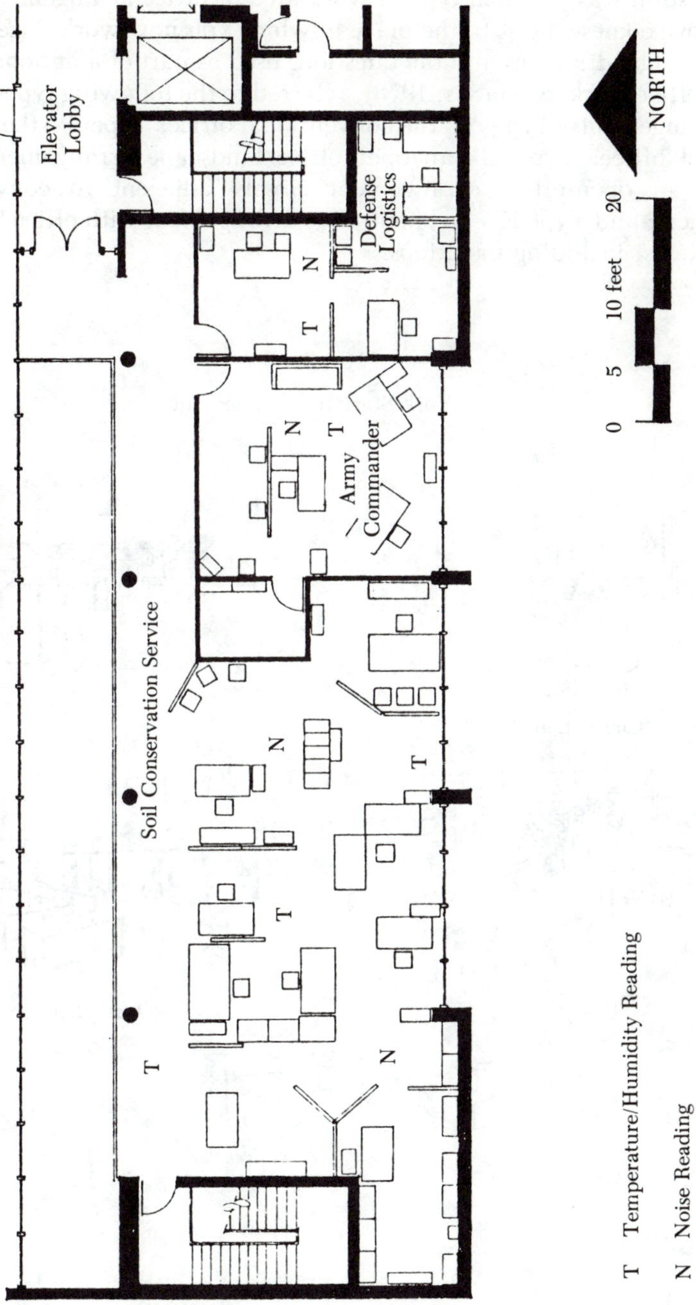

Elevator Lobby

Defense Logistics

Army Commander

Soil Conservation Service

NORTH

0 5 10 feet 20

T Temperature/Humidity Reading

N Noise Reading

question was, "Which type of work area or office arrangement shown above comes closest to the place in which you now work?" (see Figure 7.7). The drawings and the question, used as part of a national survey of office workers (Harris, 1978), referred to the following types of work arrangements: Type A—the conventional office; Type B—the clerical-pool office; Type C—an open-office landscape arrangement; Type D—an open-office arrangement closely adjacent to conventional offices; and Type E—an open-office arrangement with office furniture systems, including partitions.

FIGURE 7.7

Work Station Arrangements

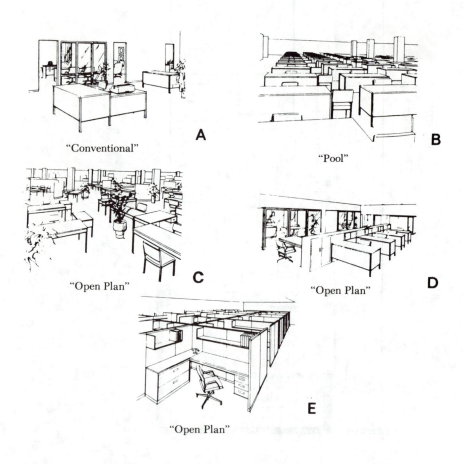

"Conventional"    A

"Pool"    B

"Open Plan"    C

"Open Plan"    D

"Open Plan"    E

*More than one-third of the work stations in the building were in pool office arrangements. In the Social Security Administration, nine out of ten employees worked in this type of setting.*

The first part of Table 7.1 shows how these arrangements are distributed in the Ann Arbor Federal Building. For the most part, the federal employees work in open-office or pool arrangements.[2] More than eight in ten selected the Types B through E drawings. The Type E arrangement was indicated most often (45 percent), particularly by people who worked in HCRS and IRS, while the clerical-pool arrangement was characteristic of the Social Security Administration. The military recruiters and employees of the small agencies were most likely to indicate they worked in a conventional office.

Federal employees were also asked, "Which type of work area comes closest to the place you worked in before coming to the Federal Building?" A significant number selected each of the five types of office arrangements. One-third said they had previously worked in a conventional office, 17 percent indicated they had come from a pool arrangement, and half had worked in an open office. Clearly, for people in most agencies, the move to the new building entailed a shift away from the conventional office to a pool or open-office arrangement.

The uniqueness of the Ann Arbor Federal Building can be seen in Table 7.2, which shows comparative data covering the national sample of office workers, including those employed by various units of government. While most office workers nationally said they worked in a conventional office, only 16 percent of the employees in the Ann Arbor Federal Building responded in this manner. By the same token, employees in the building were less likely than national respondents to work in a pool arrangement but were much more likely to report working in the Types C and E open-office arrangements.

In our second approach to determining work arrangements, members of the research team systematically observed and classified each individual work station in the building. Postal carrels and service counters were considered industrial work stations. Five classes of office work stations were identified: the conventional office used by one person; the shared conventional office; an open-office arrangement with partitioned areas occupied by one person; an open-office arrangement with partitioned spaces that were shared; and a work station in a pool arrangement. The second half of Table 7.1 presents the distribution of work stations according to these classes for each agency. Once again, the vast majority of work stations (75 percent) were classified as being in either an open or pool arrangement. Just 6 percent were considered enclosed conventional offices. It should be noted that, with the exception of the small agencies, no agency had more than 18 percent of its work stations in a conventional setting.

TABLE 7.1

Current and Previous Office Arrangement and Current Type of Work Station, by Agency
(Percentage Distribution)

| Office/Work Station | All | Agency | | | | | | |
| --- | --- | --- | --- | --- | --- | --- | --- | --- |
| | | Post Office | IRS | Military Recruiters | HCRS | Social Security | Weather Service | Small Agencies |
| Current Office Arrangement (Perceived)[a] | | | | | | | | |
| Type A "Conventional" | 18 | — | 2 | 73 | 6 | 12 | 20 | 40 |
| Type B "Pool" | 13 | — | 2 | — | — | 52 | — | 12 |
| Type C "Open Plan" | 20 | — | 43 | 27 | 9 | 30 | 20 | 8 |
| Type D "Open Plan" | 4 | — | 4 | — | — | 3 | 7 | 20 |
| Type E "Open Plan" | 45 | — | 49 | — | 85 | 3 | 53 | 20 |
| Total | 100 | — | 100 | 100 | 100 | 100 | 100 | 100 |
| Number of respondents | 177 | — | 47 | 11 | 46 | 33 | 15 | 25 |
| Previous Office Arrangement (Perceived)[b] | | | | | | | | |
| Type A "Conventional" | 33 | — | 18 | 70 | 44 | 19 | 8 | 60 |
| Type B "Pool" | 17 | — | 23 | 10 | — | 34 | 8 | 20 |
| Type C "Open Plan" | 20 | — | 23 | 10 | 10 | 41 | 23 | 8 |
| Type D "Open Plan" | 13 | — | 11 | — | 14 | 6 | 46 | 12 |
| Type E "Open Plan" | 17 | — | 25 | 10 | 32 | — | 15 | — |
| Total | 100 | — | 100 | 100 | 100 | 100 | 100 | 100 |
| Number of respondents | 165 | — | 44 | 10 | 41 | 32 | 13 | 25 |

TABLE 7.1 (Continued)

| Office/Work Station | All | Agency | | | | | | |
| --- | --- | --- | --- | --- | --- | --- | --- | --- |
| | | Post Office | IRS | Military Recruiters | HCRS | Social Security | Weather Service | Small Agencies |
| Current Work Station (Observed)[c] | | | | | | | | |
| Conventional: private | 3 | 2 | 2 | – | – | 7 | 6 | 8 |
| Conventional: shared | 3 | – | – | – | – | – | 12 | 28 |
| Open with partitions: private | 27 | – | 36 | 7 | 71 | 3 | 23 | 28 |
| Open with partitions: shared | 10 | – | 31 | – | 14 | – | – | – |
| Pool | 38 | – | 31 | 93 | 15 | 90 | 59 | 36 |
| Postal carrel/counter | 19 | 96 | – | – | – | – | – | – |
| Total | 100 | 100 | 100 | 100 | 100 | 100 | 100 | 100 |
| Number of respondents | 264 | 51 | 64 | 15 | 52 | 40 | 17 | 25 |

[a] Current office arrangement is based on the employee's selection of the drawing which best represents his or her work environment. The question was not asked of Post Office employees.

[b] Previous office arrangement is based on the employee's selection of the drawing which best represents the work environment prior to the move to the Federal Building. Besides the postal workers who were not asked the question, 11 respondents noted this was their first job and therefore had no previous office.

[c] Current work station designations were made by trained observers.

TABLE 7.2

Current Office Arrangement,
Federal Building Employees and National Data[a]
(Percentage Distribution)

| Current Office Arrangement (Perceived) | Federal Building Employees | National Data Employees[b] | |
|---|---|---|---|
| | | Government | All |
| Type A "Conventional" | 16 | 49 | 47 |
| Type B "Pool" | 12 | 21 | 17 |
| Type C "Open Plan" | 24 | 8 | 10 |
| Type D "Open Plan" | 5 | 8 | 12 |
| Type E "Open Plan" | 43 | 14 | 14 |
| Total | 100 | 100 | 100 |
| Number of respondents | 177 | 152 | 870 |

[a] The national data were generated as part of a study conducted by Louis Harris and Associates for the Steelcase Corporation (1978).

[b] The sample size for federal, state, and local governmental workers was 178. Twenty-six respondents either did not answer or said that none of the drawings described the office area in which they worked. The total sample of all office workers was 1,047, with 177 either not answering or reporting that the drawings did not represent their situation.

The research team and the employees themselves did not reach perfect agreement on the classification of work station type. As shown in Table 7.3, only one-third of the work stations classified by our observers as conventional offices were described that way by federal workers. Most of these persons indicated that one of the open-plan arrangements best represented their work situation. Similarly, in only 28 percent of the work stations we classified as a pool arrangement did the occupants select the appropriate drawing. On the other hand, 94 percent of the open-office work stations were described by employees in that manner. We suspect that this lack of association was due in part to the respondents' inability to clearly identify from the drawings the office environment most closely associated with their work situation. For example, a secretary working in an open environment adjacent to predominantly conventional offices might select the Type A drawing, intended to represent the conventional office. Similarly, people in the few private conventional offices may have had a unique arrangement relative to others in their agency but selected the drawing showing the predominant arrangement around them — the open or pool office.

Another reason for the discrepancy could have been the wording of the question, "Which type of work area or office arrangement shown above comes closest to the place in which *you now work*?"[3] In respond-

TABLE 7.3

Comparison of Respondents' Reports about Their
Office Arrangement and Observed Work Station
(Percentage Distribution)

| Current Office Arrangement (Perceived) | Current Work Station (Observed) | | | |
|---|---|---|---|---|
| | All | Conventional | Pool | Open Plan |
| Type A "Conventional" | 19 | 31 | 34 | 6 |
| Type B "Pool" | 13 | 31 | 28 | — |
| Type C, D, E "Open Plan" | 68 | 38 | 38 | 94 |
| Total | 100 | 100 | 100 | 100 |
| Number of respondents | 152 | 13 | 58 | 81 |

ing, people may have been thinking about either their agency's overall arrangement or their particular work situation within the agency. The observed classification, on the other hand, was intended to precisely describe the particular workspace of each employee. Finally, it is possible that some respondents did not carefully consider the drawings when making their choice. With these considerations in mind, the analyses presented in this chapter rely mostly on the work station classifications as we observed them.

Our third approach to examining workspace arrangements was highly impressionistic and involved visits to selected agencies by members of the research team. In each agency, observers noted the formality and organization of furnishings and equipment. Opportunities for and the amount of personalization at the work stations were noted, as were ambient environmental conditions. These observations are summarized in Table 7.4 and will be discussed later in this chapter.

*Size and density of work stations.* The size and density of each work station was measured using plans and drawings showing furniture arrangements. In each conventional office occupied by one individual, the size of the work station was determined by the square footage of the area bounded by walls. If two people occupied a conventional office, half of that square footage was ascribed to each work station. In open- and pool-office arrangements with multiple workers, the area of each work station was limited to that containing the individual's furniture and equipment and a space three feet beyond them, unless that space infringed upon the space of the neighboring work area. In that instance, the space was defined by half the distance to the nearest piece of furniture or equipment.

TABLE 7.4

Impressionistic Observations of Work Environment
and Ambient Conditions in Selected Agencies — October/November 1979

| Observations | Post Office | IRS | Military Recruiters | HCRS | Social Security | Weather Service |
|---|---|---|---|---|---|---|
| Spatial Arrangement | | | | | | |
| Organization of furniture and equipment | well organized | well organized and formal | fairly well organized/spacious | fairly well organized | fairly well organized/informal | poorly organized/ crowded |
| Opportunity for individual personalization of space | high-carriers low-clericals | high | average | high | limited | limited |
| Amount of individual personalization of space | high | low | varies | average | varies | low |
| Ambient Environmental Conditions | | | | | | |
| Noise level | high-morning | low/average | high/average | high/average | high/average | high |
| Light level | average | average/low | high | variable | high | variable |
| Temperature | low | average | low | average | average | high |

The density of each work station was measured by counting the number of surrounding work stations within a 400 square-foot area whose centroid was the center of the desk or work surface.[4]

For the entire building, the average work station included 71 square feet, with a standard deviation of 37 square feet. Actual sizes ranged from 24 to 372 square feet. The average work station density was 3.7 workers per 400 square feet; actual densities ranged from one to nine workers.[5]

In a building with such diverse activity, we would expect considerable variation among the separate agencies in work station size. As shown in Table 7.5, work stations did indeed vary, with the most spacious and least crowded found in HCRS. Work stations in the Post Office and, to a lesser extent, IRS and Social Security, were among the smallest and had the highest density. Although work stations in the Weather Service were relatively large, representing a low-density situation for its employees, it should be remembered that large amounts of equipment and high levels of personnel activity in the agency create an appearance of crowding.

*Furnishings and equipment.* It was our original plan to prepare a detailed inventory of the furnishings and equipment at each of the 265 work stations in the building; we later realized that conducting such an inventory would require considerably more manpower than we had available. Thus, data were obtained for only four characteristics: presence or absence of task lighting; type of chair; number and location of electrical and communication outlets; and the extent to which individuals personalized their work space. We had expected that variations in these characteristics would be associated with people's feelings about their work station. For instance, workers with a contoured, padded chair and plants or other desk paraphernalia, one would expect, would be more satisfied with their work station than those with a standard government-issued chair or without personal belongings on their desk.

Only six percent of the work stations in the building were equipped with task lighting. These supplementary fixtures were most likely to be used by a few employees of the HCRS, the Social Security Administration, and the Weather Service. We were told that task lighting was once used in the Post Office, but supplemental overhead lights were installed over the postal carriers' carrels within the first year of occupancy.

Work station information is presented in Table 7.6 on type of chair, communications and electrical outlets, and the degree to which the work station was personalized. Among the five classes of chairs, most were contoured and had padding, armrests, swivels, and rollers. Only

TABLE 7.5

Size and Density of Individual Work Stations, by Agency
(Mean Measures)

| Work Station Condition | All | Agency | | | | | | | |
| | | Post Office | IRS | Military Recruiters | HCRS | Social Security | Weather Service | Small Agencies |
| --- | --- | --- | --- | --- | --- | --- | --- | --- |
| Amount of work space (square feet) | 71 | 46 | 66 | 81 | 94 | 70 | 83 | 79 |
| Work space density (workers per 400 square feet) | 3.7 | 4.2 | 5.0 | 2.5 | 2.8 | 4.4 | 2.5 | 1.7 |
| Number of work stations | 265 | 51 | 64 | 15 | 53 | 40 | 17 | 25 |

TABLE 7.6

Work Station Furnishings/Equipment, by Agency
(Percentage Distribution)

| Work Station Furnishings and Equipment | All | Post Office | IRS | Military Recruiters | Agency HCRS | Social Security | Weather Service | Small Agencies |
|---|---|---|---|---|---|---|---|---|
| Type of Chair | | | | | | | | |
| Contour chair with full padding, armrests, swivel and rollers | 58 | – | 100 | 80 | 59 | 47 | 59 | 76 |
| Standard government-issued chair with padding, swivel, roller, but no armrests | 17 | – | – | 13 | 23 | 47 | 41 | 16 |
| Standard government-issued chair with padding, but no swivel or rollers | 3 | 2 | – | 7 | 7 | 3 | – | 8 |
| Stool | 20 | 98 | – | – | 4 | – | – | – |
| Other | 2 | – | – | – | 7 | 3 | – | – |
| Total | 100 | 100 | 100 | 100 | 100 | 100 | 100 | 100 |

| | | | | | | | | |
|---|---|---|---|---|---|---|---|---|
| **Communications and Electrical Outlets** | | | | | | | | |
| Work station directly over or touching outlet | 53 | — | 33 | 100 | 79 | 65 | 88 | 84 |
| Work station with electrical/communication connection but having extension cord | 15 | 2 | 22 | — | 17 | 32 | 6 | 12 |
| Work station does not have electrical/ communications service | 32 | 98 | 45 | — | 4 | 3 | 6 | 4 |
| Total | 100 | 100 | 100 | 100 | 100 | 100 | 100 | 100 |
| Number of work stations | 265 | 51 | 64 | 15 | 53 | 40 | 17 | 25 |
| **Number of Personal Objects[a]** | | | | | | | | |
| None | 55 | 66 | 79 | 39 | 38 | 42 | 75 | 35 |
| 1 or 2 objects | 25 | 27 | 15 | 46 | 26 | 22 | 25 | 39 |
| 3 or more objects | 20 | 7 | 6 | 15 | 36 | 36 | — | 26 |
| Total | 100 | 100 | 100 | 100 | 100 | 100 | 100 | 100 |
| Number of work stations[b] | 220 | 44 | 48 | 13 | 47 | 33 | 12 | 23 |

[a] For each work station or desk, the number of objects of personal nature was recorded. Objects included floor or desk plants, photographs, drawings, posters or other wall hangings, and desk paraphernalia such as pictures, radios, clocks, or other personal belongings.

[b] Data on personalization cover only those work stations for which employee data are available.

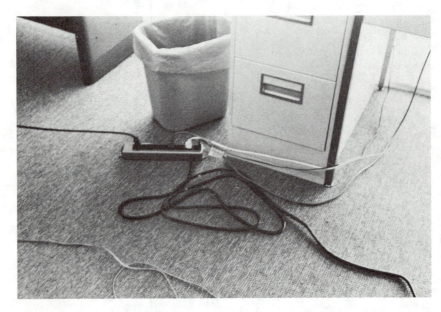

*One out of every five workers in the Internal Revenue Service and HCRS needed extension cords at their work stations. One-third of the workers in one agency had this arrangement. The number and location of electrical outlets received poor ratings by most office workers.*

three percent of the work stations had a standard government-issued chair without swivels or rollers.

More than 50 percent of the work stations had both convenient and accessible electrical and communication outlets; 15 percent had to use extension cords in order to be functionally operative. The remaining third of the work stations, mostly in the Post Office and in IRS, had no electrical or communication service whatsoever.

Although our impressionistic observations gave us some indication of the extent to which agencies were personalized, a more precise measure was needed to cover individual work stations. Accordingly, objects of a personal nature were counted for each occupied work area in the building. The data presented in the last part of Table 7.6 cover only work stations for which questionnaires were available, indicating that these places were in fact occupied. Among the 220 occupied work stations, nearly half had at least one object of a personal nature in close proximity, including photographs, drawings, posters or other wall hangings, plants, and other desk paraphernalia such as pictures, radios, or clocks. Most likely to personalize their space were people in the small agencies, in HCRS, and in Social Security and the military recruiters. The fewest objects were found at the work stations in the Weather Service and in IRS, where employees were restricted in their ability to personalize their environment. In the Weather Service, the extensive equipment and generally crowded conditions precluded employees from introducing personal objects. Security precautions in IRS required a clean-desk policy involving the removal of personal belongings from the work surface at the end of each day. Furthermore, both agencies contained a significant number of work stations that were shared.

*Accessibility.* Environmental measurements also considered how far employees were located from selected attributes considered to be either amenities or sources of personal distraction. Using floor plans for each agency, we recorded functional distances between each work station and the main agency entrance and, where such attributes were present, between the work station and the nearest window, coffee station, and lightwell.

Employee work stations were located, on average, 51 feet from the agency entrances (see Table 7.7). The shortest distance between a work station and an entrance was five feet, while the furthest distance was 125 feet. Where windows existed, the distance from the work station to the nearest window averaged 33 feet, with a range of 2 to 103 feet. For other measures as well, the ranges varied greatly, indicating the diversity of space assigned to agencies occupying the building. This

TABLE 7.7

Distances between Desks/Work Stations
and Selected Building/Agency Attributes[a]
(Feet)

| Distance between Desk/ Work Station and: | Mean | Standard Deviation | Range | Number of Desks/Work Stations Having Attributes |
|---|---|---|---|---|
| Main agency entrance | 51 | 27 | 5–125 | 265 |
| Nearest window | 33 | 23 | 2–103 | 138[b] |
| Lightwell above | 22 | 12 | 1– 61 | 134[c] |
| Lightwell below | 36 | 26 | 3– 94 | 60[d] |
| Agency's coffee station | 65 | 35 | 8–176 | 256[e] |

[a] All distances were measured from the center of the desk or work station to the attribute. For main agency entrance and the coffee pot, we considered the functional distance, while the distances to windows and lightwells were straight-line distances.

[b] There were 127 desks/work stations in either agencies with no windows or in agencies where a window exists but in an enclosed space with no glass partitions.

[c] There were 131 desks/work stations in either agencies with no lightwell above or in agencies having a lightwell but in an enclosed space with no glass partition.

[d] There were 205 desks/work stations in either agencies having no lightwell below or in agencies with a lightwell below but in an enclosed space with no glass partition.

[e] There were 9 desks/work stations in agencies having no coffee station.


diversity is reflected by the data in Table 7.8. The military recruiters and employees in the Weather Service and in the small agencies were closest to these amenities, while HCRS, IRS, and Social Security employees were the furthest away. In a later section of this chapter, these distances for individual workers will be examined relative to perceptions and evaluations of the work environment and its attributes.

*Lighting conditions.* Measurements of lighting conditions at the work stations were made using indirect and direct methods. The indirect method involved examining floor plans and describing the condition of natural light and glare at each work station. Five classes of natural light conditions were identified; these considered the extent to which the work station was located in a conventional or open office and the relationship of that office to a window or lightwell. The glare condition considered the orientation of the work station to the window or lightwell. The third measure involved a meter reading of direct light taken at the surface of each work station. The distributions for these lighting conditions in each agency is shown in Table 7.9.

The first part of the table indicates that nearly two-thirds of the work stations were situated in offices without access to natural lighting

TABLE 7.8.

Distances between Desks/Work Stations and Selected Building/Agency Attributes
(Mean Distance in Feet)[a]

| Distance between Work Station and: | All | | Agency | | | | | | |
|---|---|---|---|---|---|---|---|---|---|
| | | Post Office | IRS | Military Recruiters | HCRS | Social Security | Weather Service | Small Agencies |
| Main agency entrance | 51(265) | 40(51) | 63(64) | 17(15) | 68(53) | 62(40) | 31(17) | 27(25) |
| Nearest window | 33(138) | – | 33(25) | 16(13) | 54(43) | 23(37) | 15(13) | 10( 7) |
| Lightwell above | 22(134) | – | 23(63) | – | 20(38) | 21(33) | – | – |
| Lightwell below | 36( 60) | – | – | – | 46(43) | – | 10(13) | 17( 4) |
| Agency's coffee station | 65(256) | 66(51) | 74(64) | 20(13) | 86(53) | 61(39) | 54(17) | 26(19) |

[a] Numbers in parentheses represent number of desks/work stations and are the basis for the mean distance measures.

TABLE 7.9

Light Conditions, by Agency
(Percentage Distribution)

| | | | | | Agency | | | |
|---|---|---|---|---|---|---|---|---|
| Light Conditions | All | Post Office[a] | IRS | Military Recruiters | HCRS | Social Security | Weather Service | Small Agencies |
| Natural Light Condition | | | | | | | | |
| Closed office/no external natural light | 29 | 100 | 2 | – | – | 8 | 18 | 72 |
| Closed office/with external natural light | 1 | – | – | – | – | – | – | 12 |
| Open office/more than 10 feet from lightwell or more than 20 feet from window | 33 | – | 58 | 33 | 62 | 25 | 12 | – |
| Open office/within 20 feet of window | 19 | – | 3 | 67 | 11 | 45 | 70 | 16 |
| Open office/within 10 feet of lightwell | 18 | – | 37 | – | 27 | 22 | – | – |
| Total | 100 | 100 | 100 | 100 | 100 | 100 | 100 | 100 |
| Number of respondents | 265 | 51 | 64 | 15 | 53 | 40 | 17 | 25 |
| Glare Condition | | | | | | | | |
| Within 20 feet of north window and facing south | 6 | – | – | – | 6 | 25 | – | – |

| | | | | | | | | |
|---|---|---|---|---|---|---|---|---|
| Within 10 feet of lightwell and facing away from well | 6 | — | 13 | — | 7 | — | — | — |
| Within 20 feet of south, east or west window | 9 | — | 3 | 67 | — | — | — | 28 |
| More than 20 feet from window or more than 10 feet from lightwell | 53 | — | 59 | 33 | 64 | 35 | 29 | 72 |
| Within 20 feet of north window and facing north | 1 | — | — | — | 2 | 5 | — | — |
| Within 20 feet of north window and facing east or west | 10 | — | — | — | 4 | 15 | 71 | — |
| Within 10 feet of lightwell and facing lightwell | 15 | — | 25 | — | 17 | 20 | — | — |
| Total | 100 | — | 100 | 100 | 100 | 100 | 100 | 100 |
| Number of respondents | 214 | — | 64 | 15 | 53 | 40 | 17 | 25 |
| Light Level (foot candles) | | | | | | | | |
| 30 or less | 4 | — | 3 | 36 | 4 | — | — | 4 |
| 31–45 | 20 | 5 | 46 | 18 | 24 | 6 | 42 | 13 |
| 46–60 | 29 | 28 | 36 | 46 | 35 | 3 | 50 | 26 |
| 61–75 | 26 | 32 | 9 | — | 28 | 50 | — | 31 |
| 76–90 | 12 | 23 | 6 | — | 7 | 16 | — | 13 |
| More than 90 | 9 | 12 | — | — | 2 | 25 | 8 | 13 |
| Total | 100 | 100 | 100 | 100 | 100 | 100 | 100 | 100 |
| Number of work stations | 200 | 43 | 33 | 11 | 46 | 32 | 12 | 23 |
| Mean foot candles | 59 | 68 | 46 | 38 | 56 | 79 | 58 | 65 |

[a] Glare condition was not determined for employees of the Post Office.

(categories 1 and 3), while only one percent were situated in private offices with natural light. IRS and HCRS, because of their expansive open-office arrangements, had the highest percentage of work stations away from natural light sources. A significant proportion of work stations occupied by the military recruiters and by Social Security and Weather Service employees were in close proximity to natural light sources.

The work station at which conditions were made most deleterious because of glare were those adjacent to north windows or overhead lightwells (first, second, and last three glare condition classes). More than one-third of the work stations were characterized by one of these three classes, including two-thirds of the work stations in the Social Security Administration. Work stations in the small agencies and in the military recruiters' area all had more favorable glare conditions. In the last part of Table 7.9, the percentage distributions and means covering actual foot candles are shown. These data are based on an average of two and, in some cases, three readings taken at work stations. For the building as a whole, light levels averaged 59 foot candles, with ranges from 19 to 114.

Although they were completely isolated from any sources of natural light, the work stations in the Post Office had relatively high light levels (68 foot candles), and the work stations in the Social Security Administration, with a full wall of glass along the north, had the brightest lighting (79 foot candles). Those with the lowest lighting levels were the military recruiters — whose work stations averaged 38 foot candles, and the IRS employees — who had an average of 46 foot candles at their work stations.

*Ambient environmental conditions.* As we noted earlier, impressionistic observations of ambient environmental conditions were supplemented with precise measures taken in zones within agencies rather than at the individual work stations.[6] Temperature, relative humidity, and noise readings were found to be fairly uniform throughout large areas of the building. In Tables 7.10 and 7.11, the average readings covering these conditions are shown for the entire building and for each agency.

Shortly before the employee questionnaires were administered, the mechanical system was modified to alleviate problems experienced in the building during its first year and a half of occupancy. Several weeks later, measures of the environmental conditions were made. The relatively stable temperatures and humidity levels across agencies would indicate that the modifications had been successful. However, it should be noted that temperature readings were considerably higher

TABLE 7.10

Ambient Environmental Conditions
(265 Work Stations)[a]

| Ambient Condition | Measurement Unit | Mean | Standard Deviation | Range |
|---|---|---|---|---|
| Temperature | Degrees Farenheit (°F) | 74 | 1.8 | 71–83 |
| Relative humidity | Percentage (%) | 32 | 3.7 | 22–43 |
| Noise criteria | NC | 47 | 7.5 | 38–65 |
| Noise frequency | Hertz (Hz) | 602 | 425 | 250–4000 |
| Noise intensity | "A" weighted decibels (dBA) | 50 | 6.7 | 42–68 |

[a] Unlike lighting and other environmental measures taken at each work station in the building, readings for ambient environmental conditions were taken at selected locations within each agency.

(for the winter months) than required under federal government guidelines. Particularly high readings were found in the Weather Service area, where considerable heat was generated by the weather forecasting equipment.

Machinery and equipment were also responsible for the relatively high noise levels found in the Weather Service and in the Post Office. The movement of carts and people during the morning hours emitted high frequency sounds in the Post Office.

**Evaluating the Work Environment**

Following the questions designed to gauge employees' use of the building and their thoughts about its architectural quality, a series of evaluative questions were asked about the agency and the individual's workspace. The last question in the series was, "Overall, how satisfied are you with your work station?" More than one-third of the Federal Building workers indicated they were either not very or not at all satisfied. The remainder reported that they were very satisfied or fairly satisfied with their work stations.[7] Highest ratings were given by people in the small agencies, in the Weather Service, and by the military recruiters. Employees in IRS and HCRS reported the lowest levels of work station satisfaction (see Table 7.12).

Somewhat different results on sentiments toward the work station are indicated in the first part of Table 7.13, which shows how employees responded to the question, "Compared to where you worked before coming to the Federal Building, is your present work station better, worse, or the same?" Thirty percent reported that their present work stations were better, 42 percent said they were worse, and 23 percent

TABLE 7.11

Ambient Environmental Conditions, by Agency
(Mean Measures)

| Ambient Condition | All | Post Office | IRS | Military Recruiters | HCRS | Social Security | Weather Service | Small Agencies |
|---|---|---|---|---|---|---|---|---|
| | | | | | Agency | | | |
| Temperature (°F) | 74 | 73 | 73 | 75 | 74 | 73 | 78 | 76 |
| Relative humidity (%) | 32 | 35 | 35 | 31 | 29 | 32 | 29 | 26 |
| Noise criteria (NC) | 47 | 58 | 40 | 47 | 43 | 49 | 54 | 43 |
| Noise frequency (Hertz) | 602 | 735 | 377 | 500 | 618 | 500 | 1,382 | 560 |
| Noise intensity (dBA) | 50 | 59 | 44 | 51 | 47 | 51 | 59 | 47 |

TABLE 7.12

Satisfaction with Work Station, by Agency
(Percentage Distribution)

| Work Station Satisfaction[a] | All | Agency | | | | | | |
|---|---|---|---|---|---|---|---|---|
| | | Post Office | IRS | Military Recruiters | HCRS | Social Security | Weather Service | Small Agencies[b] |
| Very satisfied | 6 | 6 | — | — | 4 | 3 | 25 | 15 |
| Fairly satisfied | 58 | 68 | 42 | 75 | 49 | 67 | 50 | 69 |
| Not very satisfied | 28 | 26 | 37 | 17 | 38 | 30 | 13 | 8 |
| Not at all satisfied | 8 | — | 21 | 8 | 9 | — | 12 | 8 |
| Total | 100 | 100 | 100 | 100 | 100 | 100 | 100 | 100 |
| Number of respondents | 235 | 53 | 48 | 12 | 47 | 33 | 16 | 26 |
| Mean satisfaction score[c] | 2.6 | 2.8 | 2.2 | 2.7 | 2.5 | 2.7 | 2.8 | 2.9 |

[a] The question was: "Overall, how satisfied are you with your work station?"

[b] Includes Department of Defense-Army Recruiting Area Commander, Defense Logistics Agency, Defense Investigative Service and Army Surgeon General; Soil Conservation Service; District Court-Probation Department; Department of Labor, Wage and Hourly Division; the Federal Bureau of Investigation; and the security guard.

[c] Responses of "very satisfied" were coded 4, "fairly satisfied" were coded 3, "not very satisfied" were coded 2, and "not at all satisfied" were coded 1.

TABLE 7.13

Comparative Evaluation of Work Station, by Agency
(Percentage Distribution)

| | | | | | Agency | | | |
|---|---|---|---|---|---|---|---|---|
| Evaluations | All | Post Office | IRS | Military Recruiters | HCRS | Social Security | Weather Service | Small Agencies |
| Comparative Work Station Evaluation[a] | | | | | | | | |
| Better | 30 | 51 | 15 | 33 | 20 | 23 | 33 | 35 |
| Worse | 42 | 9 | 64 | 33 | 62 | 58 | 33 | 27 |
| Same | 23 | 36 | 15 | 34 | 14 | 16 | 14 | 34 |
| Better and worse | 5 | 4 | 6 | — | 4 | 3 | 20 | 4 |
| Total | 100 | 100 | 100 | 100 | 100 | 100 | 100 | 100 |
| Number of respondents | 230 | 53 | 48 | 12 | 45 | 31 | 15 | 26 |
| Reasons Work Station is Better | | | | | | | | |
| More privacy; own office; quieter | 16 | 3 | 33 | 17 | 20 | 25 | 13 | 19 |
| More space, room to work | 13 | 2 | 20 | 17 | 13 | 25 | 13 | 25 |
| Cleaner; better upkeep | 12 | 21 | 7 | 17 | — | — | 8 | 13 |
| More attractive; more modern | 11 | 14 | 13 | — | 13 | — | 13 | 6 |
| Better building, security | 9 | 16 | — | 16 | — | — | 13 | 6 |

| | | | | | | | | |
|---|---|---|---|---|---|---|---|---|
| Better furniture, storage | 8 | – | 7 | – | 34 | 13 | 20 | – |
| Good view; better lighting | 8 | 9 | 7 | – | – | 25 | – | 13 |
| Better heating, ventilation | 6 | 14 | – | – | – | – | – | 6 |
| Other | 17 | 21 | 13 | 33 | 20 | 12 | 20 | 12 |
| Total | 100 | 100 | 100 | 100 | 100 | 100 | 100 | 100 |
| Total mentions | 117 | 44 | 15 | 6 | 15 | 8 | 15 | 16 |
| Number of respondents | 76 | 27 | 10 | 4 | 11 | 7 | 8 | 9 |
| **Reasons Work Station is Worse** | | | | | | | | |
| Less space, work area; must share space | 18 | 10 | 29 | 28 | 8 | 18 | 8 | 36 |
| Less privacy | 17 | 30 | 20 | 44 | 21 | 6 | 8 | 9 |
| Poor heating, ventilation | 15 | 20 | 8 | 14 | 13 | 26 | 41 | 9 |
| No view; no windows | 11 | 20 | 5 | – | 19 | 9 | – | 27 |
| Noise; other distractions | 9 | 10 | 3 | – | 16 | 9 | 8 | – |
| Poor storage | 5 | – | 10 | – | 2 | 3 | 8 | 9 |
| Other | 25 | 10 | 25 | 14 | 21 | 29 | 27 | 10 |
| Total | 100 | 100 | 100 | 100 | 100 | 100 | 100 | 100 |
| Total mentions | 197 | 10 | 60 | 7 | 63 | 34 | 12 | 11 |
| Number of respondents | 107 | 7 | 31 | 4 | 30 | 19 | 8 | 8 |

a The question was: "Compared to where you worked before coming to the Federal Building, is your present work station better, worse, or the same?" Of the 239 employees who completed the questionnaire, 5 did not answer the question and 4 indicated they were not previously employed.

FIGURE 7.8

Comparative Assessment of Work Station —
Current Situation Relative to Past Situation
(Difference between respondents saying better
and respondents saying worse.)

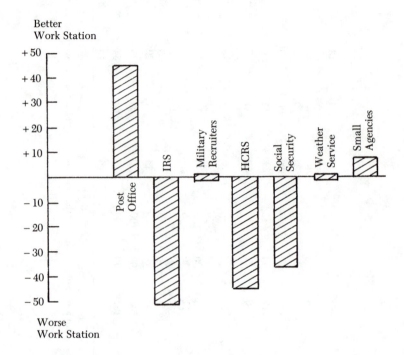

rated them the same. Most likely to report their work stations were better were the postal workers; IRS, HCRS, and Social Security employees were most likely to rate them worse. These differences are shown graphically in Figure 7.8. Clearly, most federal workers felt their work situation in terms of the immediate physical environment had deteriorated as a result of the move to the new Federal Building.

Postal workers who said their work stations were better mentioned the cleanliness and upkeep of the workspace. Employees in other agencies gave a variety of other reasons why their work stations were better. These reasons are shown in the second part of Table 7.13.

IRS personnel who said their work stations were worse were most likely to report having less space or having to share a desk with someone else. Among those in HCRS who thought their work stations

*More than four out of every ten employees thought their work areas were worse than what they had before coming to the Federal Building. Lack of privacy was often mentioned as a source of such dissatisfaction.*

were worse, mentions of less privacy were most prevalent, while Social Security employees reported having less space than before and complained about the layout and general appearance of the area around them.

An important factor influencing people's comparative work station assessments was the type of office arrangement they had relative to what they had prior to moving to the Federal Building. We found that employees who had moved from a conventional office to an open-office or pool arrangement were most inclined to say their new work stations were worse.[8] A poorer work station was least likely to be reported by those who had moved from an open-office or pool arrangement to a conventional, private office. The data in Table 7.14, nonetheless, show that a significant number of federal office employees felt their new work stations were worse, regardless of the type of change they had made.

*How do employees evalute specific work station characteristics?*

In addition to the questions dealing with their overall assessments of the work station, employees were presented with a list of specific characteristics and asked to rate each on a four-point scale ranging from excellent to poor. Among the characteristics or attributes evaluated most positively were those dealing with the access to other people and lighting. Lowest ratings were given to the outside view from the work station and the ability to conduct conversations in private. Differences in ratings of the specific attributes for employees in each agency are shown in Table 7.15 and in a set of figures in Appendix A.

Many of the items used in the list were taken from the national study of office workers (Harris, 1978) discussed earlier in this chapter. Figure 7.9 presents average ratings from the national sample on work station characteristics of three types of office arrangements; comparable ratings from Federal Building employees are depicted in Figure 7.10. In both cases, the list of characteristics tended to be rated most favorably by people occupying conventional offices. Employees working in a pool-office arrangement, on the other hand, gave the lowest ratings to specific work station characteristics.

With only one exception, ratings made by the national sample of office workers within each type of office were consistently higher than those made by the Ann Arbor Federal Building employees (see Figures 7.11, 7.12, and 7.13). There were virtually no differences in ratings of the amount of workspace between Federal Building employees and office workers from the national sample. Nonetheless, a significant number of federal employees in Ann Arbor did not like specific attributes of their assigned work stations.

TABLE 7.14

Comparative Evaluation of Work Station, by Change in Office Arrangement
(Percentage Distribution)

| Comparative Work Station Evaluation | All | Change in Office Arrangement | | | | |
|---|---|---|---|---|---|---|
| | | Conventional to Conventional | Pool to Pool | Open to Open | Conventional to Open/Pool | Open/Pool to Conventional |
| Better | 24 | 25 | 12 | 33 | 11 | 50 |
| Worse | 53 | 36 | 63 | 48 | 78 | 29 |
| Same | 23 | 39 | 25 | 19 | 11 | 21 |
| Total | 100 | 100 | 100 | 100 | 100 | 100 |
| Number of respondents | 174 | 18 | 12 | 54 | 32 | 14 |

TABLE 7.15

Ratings of Personal Work Station Characteristics, by Agency
(Mean Ratings)[a]

| Work Station Characteristics | All | Agency | | | | | | |
|---|---|---|---|---|---|---|---|---|
| | | Post Office | IRS | Military Recruiters | HCRS | Social Security | Weather Service | Small Agencies[b] |
| Access to other people | 2.9 | 2.8 | 2.6 | 3.5 | 2.7 | 2.8 | 3.1 | 3.3 |
| Location of ceiling lights in relation to work area | 2.7 | 2.8 | 2.5 | 2.7 | 2.5 | 2.7 | 2.8 | 3.3 |
| Lighting for work you do | 2.7 | 2.7 | 2.5 | 2.9 | 2.5 | 2.6 | 2.6 | 3.1 |
| Materials used for desks, tables and chairs | 2.6 | 2.5 | 2.3 | 2.7 | 3.0 | 2.8 | 2.9 | 2.7 |
| Amount of space | 2.6 | 2.4 | 2.3 | 2.3 | 3.1 | 2.3 | 2.5 | 3.0 |

TABLE 7.15 (Continued)

| Work Station Characteristics | All | Agency | | | | | | |
| --- | --- | --- | --- | --- | --- | --- | --- | --- |
| | | Post Office | IRS | Military Recruiters | HCRS | Social Security | Weather Service | Small Agencies[b] |
| Comfort of your chair | 2.5 | 1.7 | 2.7 | 2.5 | 2.6 | 2.9 | 3.0 | 2.3 |
| Amount of surface area for work | 2.4 | 2.1 | 2.1 | 2.4 | 2.7 | 2.3 | 2.4 | 2.8 |
| Type of floor covering | 2.3 | 2.3 | 2.1 | 1.7 | 2.5 | 2.3 | 1.9 | 2.4 |
| Color of walls and partitions | 2.2 | 2.5 | 2.0 | 1.7 | 2.0 | 2.3 | 2.6 | 2.6 |
| Air quality | 2.1 | 2.2 | 2.1 | 2.6 | 1.7 | 1.9 | 2.2 | 2.3 |
| Style of furniture | 2.1 | 1.8 | 2.1 | 2.1 | 2.3 | 2.3 | 2.9 | 1.7 |
| Attractiveness | 2.0 | 2.3 | 1.7 | 1.9 | 1.7 | 2.2 | 2.2 | 2.4 |
| Ventilation and air circulation | 2.0 | 2.0 | 2.0 | 2.7 | 1.8 | 1.7 | 2.0 | 1.9 |
| Overall aesthetic quality | 2.0 | 2.0 | 1.9 | 2.1 | 1.7 | 2.2 | 2.5 | 2.1 |
| Amount of space for storing | 1.9 | 1.9 | 1.7 | 1.3 | 2.0 | 1.6 | 2.0 | 2.6 |
| Wall area for hanging things | 1.9 | 1.7 | 1.6 | 3.0 | 2.1 | 1.7 | 2.1 | 2.6 |
| Heating | 1.9 | 2.1 | 1.8 | 2.2 | 2.0 | 1.5 | 1.7 | 1.9 |
| Number of electrical outlets | 1.8 | 2.1 | 1.5 | 2.1 | 1.7 | 1.8 | 1.7 | 1.9 |
| Visual privacy | 1.8 | 2.0 | 1.7 | 1.9 | 1.7 | 1.4 | 2.2 | 2.0 |
| Location of electrical outlets | 1.7 | 2.2 | 1.4 | 1.6 | 1.6 | 1.7 | 1.6 | 1.7 |
| Your view outside | 1.5 | 1.1 | 1.3 | 1.4 | 1.2 | 2.4 | 2.3 | 1.9 |
| Conversational privacy | 1.5 | 1.8 | 1.3 | 1.7 | 1.2 | 1.6 | 1.9 | 1.7 |
| Number of respondents | 238 | 54 | 49 | 12 | 47 | 33 | 16 | 27 |

[a] Ratings of "excellent" were coded 4, "good" were coded 3, "fair" were coded 2, and "poor" were coded 1.

[b] Includes Department of Defense-Army Recruiting Area Commander, Defense Logistics Agency, Defense Investigative Service and Army Surgeon General; Soil Conservation Service; District Court-Probation Department; Department of Labor, Wage and Hourly Division; the Federal Bureau of Investigation; and the security guard.

FIGURE 7.9

Average Ratings of Personal Work Station Characteristics
for Conventional, Pool and Open Offices
(National Sample of Office Workers)*

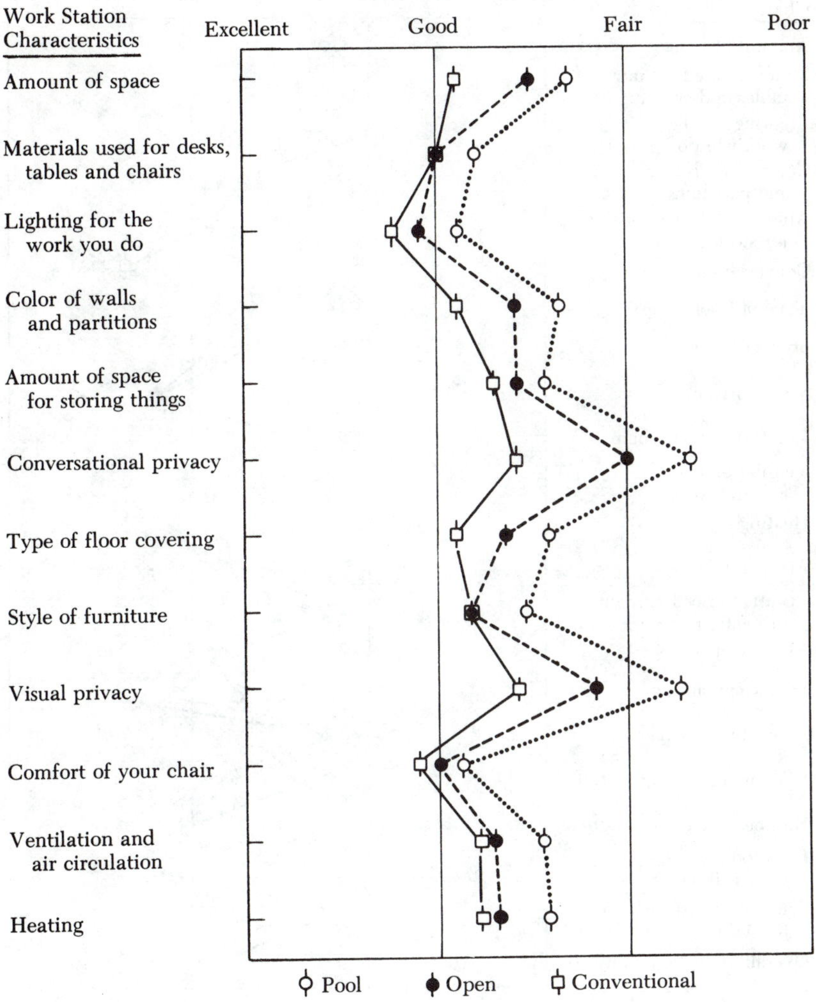

* Louis Harris and Associates, *The Steelcase National Study of Office Environments: Do They Work?*, 1978.

FIGURE 7.10

### Average Ratings of Personal Work Station Characteristics
### for Conventional, Pool and Open Offices
### (Ann Arbor Federal Building)

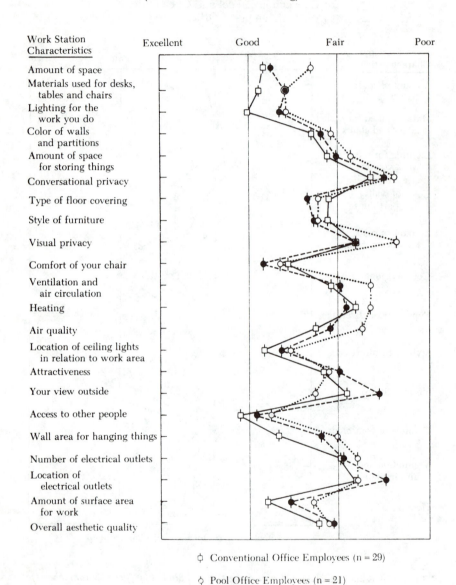

$\varphi$  Conventional Office Employees (n = 29)

$\varphi$  Pool Office Employees (n = 21)

$\blacklozenge$  Open Office Employees (n = 127)

FIGURE 7.11

Average Ratings of Personal Work Station Characteristics
for Conventional Offices
(Federal Building and National Data)

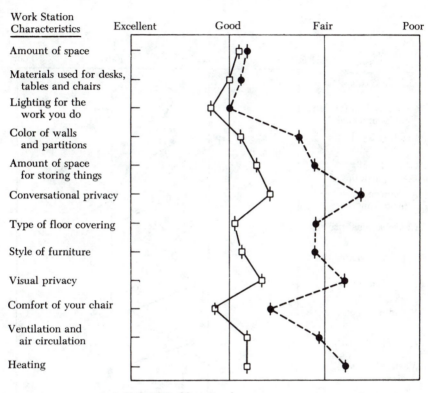

♦ Federal Building Employees

⌀ Employees from a national sample survey of office workers
conducted by Louis Harris and Associates for Steelcase, 1978.

### FIGURE 7.12

Average Ratings of Personal Work Station Characteristics for Open Offices
(Federal Building and National Data)

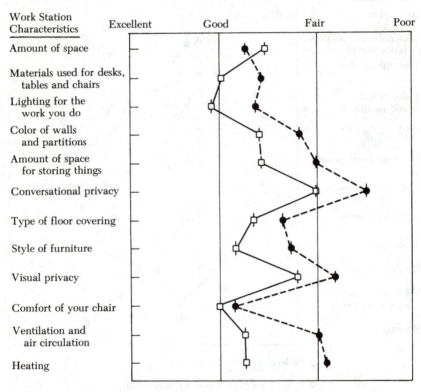

Work Station
Characteristics

Excellent          Good          Fair          Poor

Amount of space

Materials used for desks,
  tables and chairs

Lighting for the
  work you do

Color of walls
  and partitions

Amount of space
  for storing things

Conversational privacy

Type of floor covering

Style of furniture

Visual privacy

Comfort of your chair

Ventilation and
  air circulation

Heating

● Federal Building Employees

☐ Employees from a national sample survey of office workers
   conducted by Louis Harris and Associates for Steelcase, 1978.

## FIGURE 7.13

### Average Ratings of Personal Work Station Characteristics for Pool Offices
(Federal Building and National Data)

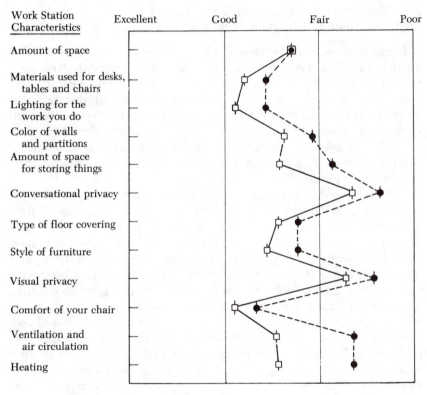

● Federal Building Employees

⌷ Employees from a national sample survey of office workers
conducted by Louis Harris and Associates for Steelcase, 1978.

The reader may note that many of the work station characteristics in the list are conceptually interrelated. In order to determine the extent to which they were statistically linked, a correlation analysis was performed and, based on the results, six indexes designed to tap worker evaluations of specific attributes of the work environment were created.[9] These indexes deal with available space, lighting at the work station, aesthetic quality of the work station, its electrical outlets, furniture, and conversational privacy.

*How do selected work station characteristics*
*relate to people's evaluations of them?*

In our discussion of the basic model in Chapter 2, we suggested that an individual's assessment of a particular environmental attribute is related to but distinct from the objective attribute itself. We also noted that, from the point of view of the environmental designer, the exploration of such relationships is an important component of the evaluation process. In this study, we examined workers' assessments of a number of specific attributes in relation to selected environmental conditions as we had measured them objectively. These relationships are shown in Figures 7.14, 7.15, and 7.16.

The first figure shows the association between the type of work station assigned to individuals and those individuals' ratings of their workspace and of visual and conversational privacy. Relationships were apparent in each case, but the type of work station was most strongly associated with the workers' feelings about the ability to carry on conversations in private. Not surprisingly, employees occupying conventional, private offices were most likely to give favorable ratings to conversational privacy. For employees who worked in an open-office or pool arrangement, we had expected their evaluations of con-versational privacy to differ depending on the types of tasks they performed. To test this, we examined the ratings for people in these offices who spent varying amounts of time on the telephone and in meeting with others at their desks. Irrespective of the type of work tasks performed, we found, employees in these settings were equally disturbed by their inability to carry on conversations in private.

People who occupied private, conventional offices were most posi-tive in their feelings about visual privacy; those in open offices or in a pool situation gave the lowest ratings. Conventional office personnel also gave higher ratings to the amount of space available to them, although people who had to share such offices gave somewhat less favorable ratings to the space than people with a private space in an open-office arrangement.

FIGURE 7.14

Relationships between Type of Work Station
and Evaluations of Space, Conversational Privacy, and Visual Privacy

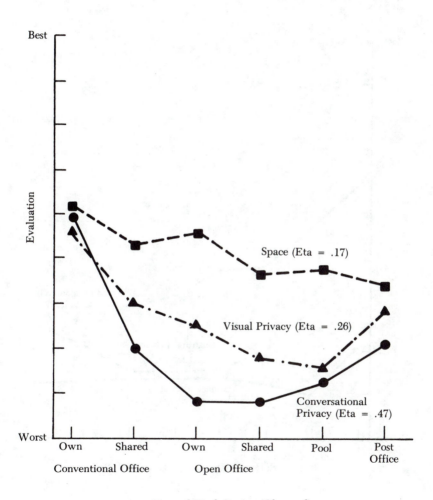

Type of Work Station (Observed)

FIGURE 7.15

Relationships between Workspace Density
and Evaluations of Space, Conversational Privacy, and Visual Privacy

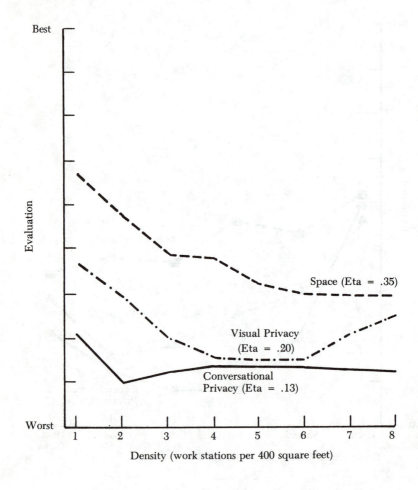

FIGURE 7.16

Relationships between the Amount of Work Space
and Evaluations of Space, Conversational Privacy, and Visual Privacy

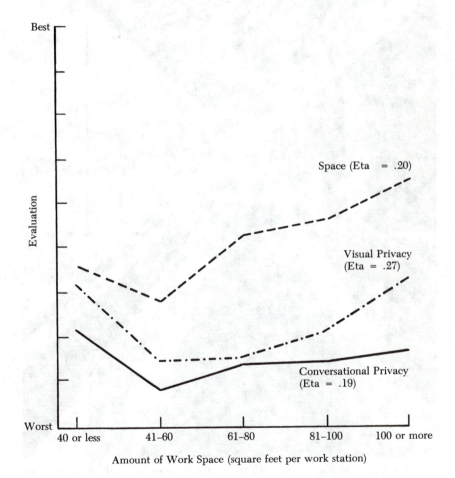

Amount of Work Space (square feet per work station)

*Employees in any agency with a lightwell opening above to another agency were the persons most likely to complain about a lack of visual and conversational privacy, irrespective of how far their own work stations were from the lightwell.*

Density of space as measured objectively was also related to people's feelings about visual privacy and the space available to them (see Figure 7.15). But, density had virtually no bearing on people's feelings about conversational privacy. Those who worked in a low-density situation (one or two work stations per 400 square feet) were just as likely to rate their conversational privacy poorly as those working in environments with more than seven work stations per 400 square feet. Finally, the amount of space in terms of square footage assigned to individuals was only moderately related to the way they felt about the space available to them or to their feelings about visual privacy. Again, actual space had no impact on ratings of conversational privacy.

Other attributes of the work station examined in relation to people's ratings were the glare condition, light intensity measured in foot candles, proximity to windows and lightwells, and noise intensity. Glare condition was weakly related to both people's feelings about their lighting situation and the extent to which they were bothered by glare. Actual light levels showed an equally weak relationship to people's ratings of the light available to them. People in work stations with 80 foot candles or more tended to rate their lighting situation more positively than those with 40 foot candles or less. But people having between 40 and 49 foot candles at their work surface rated their lighting situation most positively.

Employees who worked in areas where there was a window they could see evaluated their lighting situation lower than those in areas without windows. People who worked where there was a window they could *not* see from their work station gave the lowest ratings of all to lighting. Distance to the nearest window had no bearing on these ratings.

People in work stations beneath a lightwell gave low ratings to both visual privacy and conversational privacy. But the existence of light-wells that extended *below* an agency had no bearing on the way people viewed their privacy, nor did the distance between work stations and lightwells above or below influence ratings.

Finally, noise intensity showed a modest relationship to ratings of conversational privacy. At 60 decibels or less, people felt they had limited opportunities to conduct conversations in private. However, when the noise levels were above 60 decibels, ratings of conversational privacy improved. It would appear from these data that higher noise levels can reduce the ability to hear conversations beyond those taking place in the immediate environment and therefore contribute to a more positive (but still low) evaluation of conversational privacy.

TABLE 7.16

Overall Work Station Satisfaction,
by Selected Objective Work Station Characteristics[a]
(Mean Satisfaction Score)[b]

| Work Station Characteristics | Mean Satisfaction Score | Correlation Coefficient[c] | Number of Respondents |
|---|---|---|---|
| Amount of Workspace | | .39 | |
| 40 square feet or less | 2.8 | | 37 |
| 41–60 square feet | 2.4 | | 37 |
| 61–80 square feet | 2.9 | | 61 |
| 81–100 square feet | 2.7 | | 48 |
| More than 100 square feet | 2.0 | | 21 |
| Chair Type | | .32 | |
| Contoured chair with full padding, armrests, swivel, and rollers | 2.4 | | 111 |
| Standard government-issued chair with padding, swivel, rollers, but no armrests | 3.0 | | 41 |
| Stool | 2.3 | | 42 |
| Window Condition in Agency | | .26 | |
| No window | 2.9 | | 89 |
| Window/no visual access | 2.3 | | 13 |
| Window in agency/visual access | 2.7 | | 102 |
| Glare Condition | | .26 | |
| Within 20 feet of south, east, or west window | 2.7 | | 15 |
| More than 20 feet from window or more than 10 feet from lightwell | 2.7 | | 81 |
| Within 10 feet of lightwell and facing lightwell | 2.6 | | 23 |
| Within 20 feet of north window and facing east, north, or west | 2.7 | | 18 |
| Within 20 feet of north window and facing south | 2.6 | | 9 |
| Within 10 feet of lightwell and facing away from well | 1.8 | | 12 |
| Current Work Station | | .25 | |
| Conventional: private | 2.7 | | 8 |
| Conventional: shared | 2.7 | | 6 |
| Open: private | 2.4 | | 63 |
| Open: shared | 2.0 | | 10 |
| Pool | 2.8 | | 75 |
| Postal carrel/counter | 2.8 | | 41 |
| Lightwell above Agency | | .22 | |
| No lightwell above | 2.8 | | 89 |
| Lightwell above/no visual access | 2.5 | | 13 |
| Lightwell above/visual access | 2.5 | | 102 |
| Natural Light Condition | | .22 | |
| Closed office | 2.8 | | 67 |

TABLE 7.16 (Continued)

| Work Station Characteristics | Mean Satisfaction Score | Correlation Coefficient[c] | Number of Respondents |
|---|---|---|---|
| Natural Light Condition (continued) | | | |
| Open office/within 20 feet of window | 2.7 | | 39 |
| Open office/more than 10 feet from lightwell or more than 20 feet from window | 2.6 | | 60 |
| Open office/within 10 feet of lightwell | 2.3 | | 38 |

[a] Data are presented for only work station characteristics associated with satisfaction with the work station at the level of .20 or higher.

[b] Responses of "very satisfied" were coded 4, "fairly satisfied" were coded 3, "not very satisfied" were coded 2, and "not at all satisfied" were coded 1.

[c] The correlation coefficient is Eta.

## How do work station characteristics relate to overall work station satisfaction?

In our discussion of the conceptual model presented in Chapter 2, we suggested that the objective attributes of the environment are directly linked to people's perceptions and assessments of those attributes. At the same time, we indicated that they contribute to a global assessment or measure of satisfaction with the overall environment or place. The extent to which these indirect links exist between the objective attributes and people's satisfaction with their work stations are shown in Table 7.16. The strongest relationships are for the amount of work-space, the style of chair, window condition, and type of work-station arrangement. However, the relationships were not always in the expected direction. For example, those with more than 100 square feet of workspace were least satisfied with their work stations, while those with 40 square feet or less were most satisfied. And employees who worked in an office-pool arrangement were just as satisfied with their individual work stations as those working in conventional private offices.

People in agencies without windows expressed high levels of work station satisfaction, while persons in agencies having a window they couldn't see from their desks were least satisfied. Finally, people who sat in standard government-issued chairs were much more satisfied with their work stations than those who sat in contoured chairs with full padding. Surprisingly, no relationships were found between work station satisfaction and the density of the workspace, the distance from the work station to the nearest agency entrance, a window, or the agency's coffee station, or the extent to which individuals personalized their workspace.

While some objective attributes of the work environment were associated with the way people felt about their work stations, no doubt other factors of a more subjective nature were also important to work station satisfaction.

*Which work station evaluations are most important*
*in understanding overall work station satisfaction?*

Earlier, we showed how employees rated specific work station characteristics and how these ratings varied for people in different agencies. As implied by our conceptual model, we would expect that evaluations of some specific characteristics would contribute to people's overall work station satisfaction, but that the characteristics (and the evaluations of them) would take on greater or lesser importance depending on who was making the assessment. As a first step in examining the relative importance of the evaluations of individual characteristics to overall work station satisfaction, we considered bivariate relationships. Based on the results, seven evaluative measures were examined simultaneously in relation to satisfaction with the work station. Together these explained nearly one-fourth of the variation in the level of satisfaction expressed by respondents. The most important predictors were people's feelings about the aesthetic quality of the workspace and the amount of space available to them.[10] Least important to overall satisfaction was their evaluation of the furniture.[11]

The four most important evaluative items were then considered along with agency and job classification in a model predicting to work station satisfaction. Table 7.17 shows that these evaluative items added 19.4 percentage points to the explained variance over and above agency and job classification. Of the predictors considered in this analysis, the evaluation of the amount of space was the most important.

When four objective characteristics of the work station were added to the model, the explained variance increased to 41 percent. The actual amount of workspace was the most important predictor in the new model. What is surprising, however, is that while satisfaction increased as the amount of space increased up to 100 square feet, work station satisfaction declined for Federal Building employees with 100 square feet or more of workspace.

The relationships suggested by our model regarding how people evaluate one aspect of their work settings—the work station—do in

TABLE 7.17

Satisfaction with Work Station, Predicted by
Objective Work Station Attributes and Evaluation of Work Station Attributes
(Multiple Classification Analyses; N = 194)

| Predictors | Eta Coefficient | Beta Coefficient[a] | | | |
|---|---|---|---|---|---|
| | | Employee Characteristics Only | Employee Characteristics and Attribute Ratings | Employee Characteristics and Objective Attributes | Employee Characteristics, Objective Attributes, and Attribute Ratings |
| Employee Characteristics | | | | | |
| Agency | .28 | .34(1) | .17(5) | .20(3) | .39(2) |
| Job classification | .24 | .26(2) | .28(2) | .20(4) | .27(5) |
| Objective Attributes | | | | | |
| Amount of workspace | .39 | | | .49(1) | .50(1) |
| Chair type | .33 | | | .19(5) | .16(8) |
| Work station type | .25 | | | .25(2) | .36(3) |
| Window condition | .26 | | | .14(6) | .31(4) |
| Attribute Ratings | | | | | |
| Aesthetic quality | .38 | | .18(4) | | .15(9) |
| Space | .33 | | .29(1) | | .26(6) |
| Conversational privacy | .30 | | .15(6) | | .14(10) |
| View outside | .21 | | .19(3) | | .17(7) |
| Percentage of variance explained | | | | | |
| (adjusted multiple $R^2$) | | 11.2 | 30.6 | 25.5 | 41.1 |
| (unadjusted multiple $R^2$) | | 15.4 | 38.3 | 34.3 | 52.1 |

[a] Numbers in parentheses indicate ranking of importance.

fact hold true. Employees' satisfaction with their work stations was found to be a function of not only who they were and where they worked, but also of the level of specific environmental attributes available to them and how they evaluated these attributes.

### How do federal employees evaluate the overall ambience of their agencies?

Federal employees were asked questions about the overall space available to their agencies — that is, the office and other workspaces assigned to the organizations in which they worked. The evaluative questions were asked about the agency's appearance and its functional arrangement. Responses to the question dealing with appearance were subsequently combined with another evaluative question covering the degree to which employees thought their physical surroundings were pleasant; these two items were part of an evaluative index reflecting people's feelings about the general ambience of their agency. Responses to the two items and the composite index of agency ambience are shown in Table 7.18 for employees in different parts of the building.[12]

A significant number of federal employees were not very happy with their physical surroundings beyond the immediate work station. Twenty-five percent gave poor ratings to their agency's appearance and 42 percent considered their physical surroundings unpleasant. Those most dissatisfied with their agency's ambience were employees of the HCRS and IRS. As in the case of work station evaluations, the military recruiters and the people working in the small agencies were most content.[13]

### How distracting are selected ambient environmental conditions?

We asked the federal employees about a number of ambient conditions and the extent to which they were distracting or bothersome. These conditions dealt with noise, lighting, heating and ventilating, and the movement of people and furniture. Employees were presented with a list of 21 such conditions and were asked to indicate the degree to which each was bothersome. Responses were given on a four-point scale ranging from very bothersome to not at all bothersome. As Table 7.19 shows, the most bothersome conditions were those dealing with heating, ventilating, and noise. For example, 41 percent reported that being too hot in the summer was bothersome, and 42 percent felt the same about the building being too cold in the winter. At the other extreme, only 2 percent said that noise from the ventilating system was very bothersome, while one percent responded in this manner to the item about street noise.

TABLE 7.18

Ratings of Agency Ambience, by Agency[a]

(Percentage Distribution)

| Ratings | All | Post Office | IRS | Military Recruiters | HCRS | Social Security | Weather Service | Small Agencies |
|---|---|---|---|---|---|---|---|---|
| | | | | | Agency | | | |
| Appearance of Agency | | | | | | | | |
| Excellent | 5 | 8 | — | 25 | — | — | — | 15 |
| Pretty good | 33 | 46 | 14 | 58 | 15 | 36 | 31 | 56 |
| Fair | 37 | 40 | 44 | 9 | 40 | 40 | 50 | 18 |
| Poor | 25 | 6 | 42 | 8 | 45 | 24 | 19 | 11 |
| Total | 100 | 100 | 100 | 100 | 100 | 100 | 100 | 100 |
| Number of respondents | 234 | 50 | 49 | 12 | 47 | 33 | 16 | 27 |
| "The physical surroundings are pleasant." | | | | | | | | |
| Very true | 10 | 15 | 4 | 7 | 2 | 15 | 6 | 20 |
| Somewhat true | 48 | 53 | 37 | 75 | 42 | 40 | 62 | 64 |
| Not very true | 30 | 26 | 42 | 9 | 41 | 36 | 13 | 8 |
| Not true at all | 12 | 6 | 17 | 9 | 15 | 9 | 19 | 8 |
| Total | 100 | 100 | 100 | 100 | 100 | 100 | 100 | 100 |
| Number of respondents | 233 | 53 | 48 | 12 | 46 | 33 | 16 | 25 |

TABLE 7.18 (Continued)

| Ratings | | Agency | | | | | | | |
|---|---|---|---|---|---|---|---|---|---|
| | All | Post Office | IRS | Military Recruiters | HCRS | Social Security | Weather Service | Small Agencies |
| Agency Ambience | | | | | | | | |
| (8) Positive rating | 2 | 4 | – | – | – | – | – | 8 |
| (7) | 9 | 10 | 6 | 31 | 2 | 12 | 6 | 8 |
| (6) | 28 | 34 | 9 | 46 | 13 | 18 | 25 | 57 |
| (5) | 23 | 28 | 26 | 8 | 21 | 30 | 25 | 8 |
| (4) | 20 | 18 | 28 | – | 25 | 15 | 25 | 11 |
| (3) | 13 | 4 | 14 | 15 | 26 | 18 | 13 | – |
| (2) Negative rating | 8 | 2 | 17 | – | 13 | 6 | 6 | 8 |
| Total | 100 | 100 | 100 | 100 | 100 | 100 | 100 | 100 |
| Number of respondents | 220 | 50 | 35 | 13 | 47 | 33 | 16 | 26 |
| Mean rating[b] | 3.8 | 4.3 | 3.1 | 4.8 | 3.0 | 3.7 | 3.7 | 4.6 |

[a] Respondents were asked to evaluate the overall space available to their agencies on: "the way the *overall* space looks." They were also asked to indicate how true a number of statements were, including one on dealing with physical surroundings. The two responses were combined to create a measure of agency ambience.

[b] The higher the score, the more positive the rating.

TABLE 7.19

Agency and Work Station Distractions
(Percentage Distribution)

| Agency and Work Station Distractions | Rating | | | | Total | Number of Respondents | Mean Score[a] |
|---|---|---|---|---|---|---|---|
| | Very Bothersome | Fairly Bothersome | Not Very Bothersome | Not At All Bothersome | | | |
| Too hot in summer | 41 | 24 | 23 | 12 | 100 | 217 | 2.1 |
| Too cold in winter | 42 | 24 | 20 | 14 | 100 | 216 | 2.1 |
| Ringing telephones in own agency | 24 | 41 | 19 | 16 | 100 | 231 | 2.3 |
| Stuffy air | 26 | 26 | 19 | 29 | 100 | 227 | 2.5 |
| Conversations of others in own agency | 14 | 34 | 27 | 25 | 100 | 229 | 2.6 |
| Drafts | 30 | 16 | 24 | 30 | 100 | 217 | 2.6 |
| Too hot in winter | 23 | 18 | 24 | 35 | 100 | 207 | 2.7 |
| Noise from equipment in own agency | 11 | 31 | 31 | 27 | 100 | 228 | 2.7 |
| Too cold in summer | 19 | 15 | 20 | 46 | 100 | 210 | 2.9 |
| People walking around | 10 | 19 | 30 | 41 | 100 | 227 | 3.0 |
| Frequent rearranging of furniture | 9 | 14 | 28 | 49 | 100 | 221 | 3.2 |
| Noise from telephone in other agencies | 9 | 17 | 13 | 61 | 100 | 228 | 3.3 |
| Glare from ceiling lights | 9 | 9 | 26 | 56 | 100 | 226 | 3.3 |
| Conversations from other agencies | 9 | 11 | 11 | 69 | 100 | 228 | 3.4 |
| Noise from equipment from other agencies | 10 | 8 | 14 | 68 | 100 | 228 | 3.4 |
| Noise from ventilating system | 2 | 10 | 26 | 62 | 100 | 230 | 3.5 |
| Heat from natural sunlight | 7 | 8 | 13 | 72 | 100 | 213 | 3.5 |
| Frequent rearranging of lighting fixtures | 2 | 2 | 16 | 80 | 100 | 221 | 3.7 |
| Glare from natural sunlight | 4 | 3 | 13 | 80 | 100 | 216 | 3.7 |
| Noise from public lobby/corridors | 2 | 3 | 15 | 80 | 100 | 229 | 3.7 |
| Noise from street and parking lot | 1 | 2 | 12 | 85 | 100 | 230 | 3.8 |

[a] Mean scores were coded as follows: 1 for "very bothersome," 2 for "fairly bothersome," 3 for "not very bothersome," and 4 for "not at all bothersome." The lower the score, the more bothersome the distraction.

In considering average responses to the conditions within each agency, significant differences existed, depending on where within the building individuals worked (Table 7.20). For example, employees in the Weather Service and in the Social Security Administration were most likely to complain about it being too hot in the summer; the military recruiters and those in HCRS were least likely to complain. Least likely to be bothered by conversations from others around them were the workers in the small government units and in the Post Office. IRS, HCRS, and Social Security employees were most vocal in their complaints about the conversations of others in their own agencies.

As in the case of work station characteristics, several of the conditions presented on the list were conceptually and statistically related. For example, people who were bothered by the telephone ringing in their own agency were also bothered by the noise from nearby equipment and the conversations taking place around them; and people who complained about their space being too hot in the summer were also likely to report it too cold in the winter. Others responded in an opposite manner: those saying it was too cold in the summer tended to say the building was too hot in the winter. And the people who were bothered by others walking around them were also bothered by frequent furniture rearrangements. Several of these items were combined into indexes reflecting the multidimensional nature of ambient conditions. These indexes deal with the noise from within one's own agency, the noise from other agencies, movements, temperature overcompensation, temperature undercompensation, and air quality.[14] Average ratings on each of these distractions are shown in Table 7.21.

*What environmental conditions are likely to*
*be distracting to Federal Building employees?*

Earlier in this chapter, data were presented on relationships between people's responses to attributes of the work environment and several environmental conditions. These relationships were suggested as part of the model outlined in Chapter 2. Consideration has also been given to the manner in which these objective conditions operating within agencies are related to people's perceptions of the conditions around them (Table 7.22). One relationship is worth noting.[15]

People who were distracted by noise from other agencies were most likely to work in spaces near a lightwell, either below or above, where the noise level was about 55 decibels, and the predominant arrangement was the open office (Table 7.22).

Clearly, agency differences accounted for most of the variance in the degree to which employees were bothered by outside noise. Irrespec-

TABLE 7.20

Agency and Work Station Distractions, by Agency
(Mean Level of Distraction)[a]

| Work Station and Agency Distractions | All | Agency | | | | | | |
|---|---|---|---|---|---|---|---|---|
| | | Post Office | IRS | Military Recruiters | HCRS | Social Security | Weather Service | Small Agencies |
| Too hot in summer | 2.1 | 2.1 | 2.0 | 2.5 | 2.4 | 1.7 | 1.2 | 2.3 |
| Too cold in winter | 2.1 | 2.4 | 1.7 | 2.6 | 2.4 | 1.4 | 1.1 | 2.8 |
| Ringing telephones in own agency | 2.3 | 2.4 | 1.8 | 3.0 | 2.0 | 1.9 | 2.7 | 3.2 |
| Stuffy air | 2.5 | 3.0 | 2.5 | 2.8 | 2.1 | 2.2 | 2.5 | 2.7 |
| Conversations of others in own agency | 2.6 | 3.2 | 2.3 | 2.8 | 2.2 | 2.3 | 2.9 | 3.2 |
| Drafts | 2.6 | 2.3 | 2.7 | 3.1 | 2.6 | 2.1 | 2.5 | 3.1 |
| Too hot in winter | 2.7 | 3.0 | 3.2 | 2.3 | 2.5 | 2.4 | 2.4 | 2.6 |
| Noise from equipment in own agency | 2.7 | 2.6 | 2.9 | 3.3 | 2.5 | 2.4 | 2.1 | 3.7 |
| Too cold in summer | 2.9 | 3.8 | 3.0 | 2.1 | 2.5 | 1.9 | 3.2 | 3.5 |
| People walking around | 3.0 | 3.5 | 2.8 | 3.5 | 2.7 | 2.9 | 2.7 | 3.5 |
| Frequent rearranging of furniture | 3.2 | 3.7 | 3.0 | 3.5 | 2.8 | 2.7 | 2.7 | 4.0 |
| Noise from telephones in other agencies | 3.3 | 3.9 | 3.1 | 3.6 | 2.7 | 4.0 | 2.4 | 3.1 |
| Glare from ceiling lights | 3.3 | 3.4 | 3.0 | 3.8 | 3.1 | 3.6 | 2.6 | 3.7 |
| Conversations from other agencies | 3.4 | 4.0 | 3.3 | 3.7 | 2.4 | 4.0 | 3.6 | 3.3 |
| Noise from equipment from other agencies | 3.4 | 3.9 | 3.3 | 3.8 | 2.4 | 4.0 | 3.7 | 3.2 |
| Noise from ventilating system | 3.5 | 3.7 | 3.6 | 3.4 | 3.5 | 3.1 | 3.6 | 3.2 |
| Heat from natural sunlight | 3.5 | 3.9 | 3.6 | 3.9 | 3.7 | 2.6 | 3.6 | 3.7 |
| Frequent rearranging of lighting fixtures | 3.7 | 3.9 | 3.6 | 3.6 | 3.8 | 3.7 | 3.8 | 4.0 |
| Glare from natural sunlight | 3.7 | 3.9 | 3.6 | 3.6 | 3.9 | 3.3 | 3.4 | 4.0 |
| Noise from public lobby/corridors | 3.7 | 3.8 | 3.7 | 3.5 | 3.8 | 3.7 | 3.9 | 3.5 |
| Noise from street and parking lot | 3.8 | 3.9 | 3.6 | 3.8 | 3.9 | 3.8 | 3.7 | 4.0 |
| Number of respondents | 231 | 47 | 49 | 12 | 47 | 33 | 16 | 27 |

[a] Level of distraction is expressed as the degree to which a work station or agency condition is bothersome. Responses ranged from "very bothersome" (coded 1) to "not at all bothersome" (coded 4).

TABLE 7.21

Distractions, by Agency
(Mean Score)[a]

| | | | | Agency | | | | |
|---|---|---|---|---|---|---|---|---|
| Distraction Index[b] | All | Post Office | IRS | Military Recruiters | HCRS | Social Security | Weather Service | Small Agencies |
| Other agency noise | 1.8 | 1.1 | 1.9 | 1.5 | 2.8 | 1.0 | 2.3 | 1.8 |
| Own agency noise | 2.6 | 2.3 | 2.9 | 2.0 | 2.9 | 2.9 | 2.5 | 1.7 |
| Movement | 2.1 | 1.5 | 2.3 | 1.6 | 2.5 | 2.5 | 2.4 | 1.3 |
| Temperature overcompensation | 2.8 | 2.6 | 3.0 | 2.2 | 2.6 | 2.6 | 3.8 | 2.4 |
| Temperature undercompensation | 2.3 | 1.8 | 2.0 | 2.9 | 2.6 | 2.6 | 2.4 | 2.2 |
| Air quality | 2.5 | 2.4 | 2.4 | 2.0 | 2.9 | 2.8 | 2.4 | 2.3 |
| Number of respondents | 239 | 54 | 49 | 13 | 47 | 33 | 16 | 27 |

[a] Scores range from 1 to 4, with 1 representing low levels of distraction and 4 representing high levels of distraction.

[b] Distractions are based on responses to individual questions and combined into indexes. Individual items and their intercorrelations are shown in Appendix Table A.3.

TABLE 7.22

Perceptions of Noise from Other Agencies Predicted by Ambient Environmental Conditions
(Multiple Classification Analysis; N = 194)

| Predictor | Eta Coefficient | Beta Coefficient[a] | | |
|---|---|---|---|---|
| | | Employee Characteristics Only | Ambient Conditions Only | Employee Characteristics and Ambient Conditions |
| Employee Characteristics | | | | |
| Agency | .63 | .67(1) | | 1.20(1) |
| Job classification | .32 | .14(2) | | .14(5) |
| Ambient Conditions | | | | |
| Lightwell below | .56 | | .44(1) | .23(3) |
| Work station type | .46 | | .17(2) | .16(4) |
| Noise intensity (decibels) | .33 | | .17(3) | .13(6) |
| Lightwell above? | .29 | | .04(4) | .71(2) |
| Percentage of variance explained (adjusted multiple $R^2$) | | 37.9 | 34.7 | 39.0 |

[a] Numbers in parentheses indicate ranking of importance.

tive of the ambient conditions and the kinds of work people were doing, HCRS employees were most likely to complain about noise from
other agencies, and the military recruiters on the first floor were least
likely to complain. But even after taking into account their agency affiliation, having a lightwell opening up above them to another agency
was the most important factor associated with people's complaints.
That is, the employees in IRS and HCRS, where such a condition exists,
were more likely to complain about noise than Weather Service
employees who only had a lightwell below them. The Weather Service
was the noisiest agency in terms of our objective measures, and noise
measures in the HCRS near the lightwell opening to the Weather Service above were nearly as high.[16]

*To what extent do people's perceptions of the ambient
conditions around them contribute to their evaluations
of the overall ambience of their agencies?*

Although our measure of agency ambience has an aesthetic component to it, it also embodies other dimensions of the physical setting in
which workers perform day-to-day functions. Noise levels, temperature, humidity, and activities taking place around workers are part
and parcel of that setting. To test this proposition, we examined actual
conditions within each agency relative to people's perceptions of a
number of ambient conditions and the degree to which they found
them bothersome. Four conditions were found to be related to people's
feelings about the overall ambience of their agencies: noise from other
agencies, noise from their own agency, the movements of other people
and equipment, and the quality of the building's air. When people's
perceptions of these four ambient conditions were considered simultaneously in predicting to overall agency ambience, 30 percent of the
variance in responses was explained.

Air quality was the most important ambient condition, followed by
noise from other agencies. People most bothered by the quality of the
air around them and the noise from elsewhere in the building were
most likely to give low ratings to agency ambience. The extent to which
these assessments were related to agency ambience, irrespective of
agency or type of work, is shown in Table 7.23.[17]

The HCRS and IRS spaces were viewed most critically by their
employees. Nonetheless, perception of air quality was the best indicator of how a person assessed the overall ambience of his or her
agency.

In Chapter 6, we showed how people's assessments of the ambience
of their agency significantly contributed to their feelings about the

TABLE 7.23

Evaluation of Agency Ambience Predicted by Employee Characteristics,
Their Perceptions of Ambient Condition and Work Station Satisfaction
(Multiple Classification Analysis; N = 202)

| Predictors | Eta Coefficient | Beta Coefficient[a] | | | |
|---|---|---|---|---|---|
| | | Employee Characteristics Only | Perceptions Only | Employee Characteristics and Perceptions | Employee Characteristics, Perceptions, and Work Station Satisfaction |
| Employee Characteristics | | | | | |
| Agency | .43 | .46(1) | | .27(2) | .27(2) |
| Job classification | .24 | .15(2) | | .15(4) | .07(7) |
| Perceptions of Ambient Conditions | | | | | |
| Air quality | .39 | | .38(1) | .38(1) | .24(3) |
| Noise from other agencies | .37 | | .24(2) | .21(3) | .16(4) |
| Movements | .35 | | .15(3) | .13(5) | .11(5) |
| Noise from own agencies | .20 | | .14(4) | .09(6) | .10(6) |
| Work Station Satisfaction | .56 | | | | .43(1) |
| Percentage of variance explained (adjusted multiple $R^2$) | | 16.0 | 30.4 | 34.6 | 46.6 |

[a] Numbers in parentheses indicate ranking of importance.

architectural quality of the building. Employees who were critical of their agency's ambience were most likely to give low marks to architectural quality, while those who gave a positive rating to their agency's ambient environment were most praiseworthy of the building's architectural design. People's ratings of their immediate work setting also contributed to their feelings about the environment of the agency within which they worked. As the last part of Table 7.23 shows, work station satisfaction added one-third to the explained variance over and above the employees' characteristics and their views on specific ambient conditions. To a large extent, a person's feelings about his work station reflected his attitudes toward its aesthetic quality. But those feelings also reflected the amount and type of workspace he had and his assessments of the view from the work situation (see Table 7.16). We can conclude from this analysis that ratings of the ambient environment are a function both of people's perceptions of the physical conditions of that environment — such as temperature and noise — and the situation experienced by workers at their immediate work station.

*How do federal employees evaluate their*
*agency's functional arrangement?*

In addition to evaluating agency ambience, the federal employees were asked to rate their agency's functional arrangement or the way offices and other spaces were arranged in terms of making it easier for them to do their jobs well. Table 7.24 shows that six in ten gave only fair or poor ratings to the organization and layout of their agencies. Lowest ratings were reported by HCRS employees; the military recruiters and small agency personnel gave the most positive evaluations. The spatial organization of the agencies in the Ann Arbor Federal Building looked particularly bad when employee responses were compared with those given by office workers in response to the same question used in the national study (Harris, 1978). Only one-third of the office workers nationally gave negative ratings to the way spaces around them were arranged. But the national study also showed that government workers in general were not happy with the functional arrangement of their organizations. Forty-seven percent of the people working at all levels of government gave negative ratings to their agency's functional arrangement.

*To what extent do ambient conditions influence people's*
*ratings of the functional arrangement of their agencies?*

Ratings of several ambient environmental conditions were examined vis-à-vis people's feelings about their agency's functional arrangement.

TABLE 7.24

Ratings of the Agency Functional Arrangement, by Agency[a]
(Percentage Distribution)

| Rating of Agency Functional Arrangement | All | Agency | | | | | | |
| --- | --- | --- | --- | --- | --- | --- | --- | --- |
| | | Post Office | IRS | Military Recruiters | HCRS | Social Security | Weather Service | Small Agencies |
| (4) Excellent | 6 | 10 | 4 | 17 | — | — | — | 19 |
| (3) Pretty good | 33 | 32 | 12 | 58 | 21 | 39 | 56 | 55 |
| (2) Fair | 33 | 42 | 29 | 8 | 43 | 46 | 25 | 11 |
| (1) Poor | 28 | 16 | 55 | 17 | 36 | 15 | 19 | 15 |
| Total | 100 | 100 | 100 | 100 | 100 | 100 | 100 | 100 |
| Number of respondents | 234 | 50 | 49 | 12 | 47 | 33 | 16 | 27 |
| Mean rating[b] | 2.6 | 2.4 | 1.7 | 2.8 | 1.9 | 2.1 | 2.4 | 2.8 |

[a] Respondents were asked to evaluate the overall space available to their agencies on the following: "The way offices and other spaces are arranged in terms of making it easier for employees to get their jobs done well."
[b] Responses ranged from "excellent" (coded 4) to "poor" (coded 1).

TABLE 7.25

Evaluation of Agency's Functional Arrangement Predicted by Employee Characteristics,
Their Perceptions of Ambient Conditions and Work Station Satisfaction
(Multiple Classification Analysis; N = 202)

| Predictors | Eta Coefficient | Beta Coefficient[a] | | | |
| --- | --- | --- | --- | --- | --- |
| | | Employee Characteristics Only | Perceptions Only | Employee Characteristics and Perceptions | Employee Characteristics, Perceptions, and Work Station Satisfaction |
| Employee Characteristics | | | | | |
| Agency | .40 | .48(1) | | .32(1) | .24(2) |
| Job classification | .23 | .22(2) | | .23(2) | .17(3) |
| Perceptions of Ambient Conditions | | | | | |
| Movement | .31 | | .23(1) | .22(3) | .16(4) |
| Noise from own agency | .31 | | .23(2) | .13(4) | .07(5) |
| Work Station Satisfaction | .51 | | | | .41(1) |
| Percentage of variance explained (adjusted multiple $R^2$) | | 16.0 | 11.5 | 21.2 | 33.5 |

[a] Numbers in parentheses indicate ranking of importance.

Employees who worked close to their agency's entrance (within 40 feet) and in conventional offices gave higher ratings to the functional arrangement than those who worked more than 80 feet from the entrance and in an open office. Of the two factors, distance was somewhat more important to the ratings.

Of the ambient conditions considered to be distracting to employees, movements of people and equipment and noise from other agencies were most likely to be associated with low ratings of the functional arrangement. However, as seen in Table 7.25, these relationships were not particularly strong. Together, the two subjective measures explained slightly more than a tenth of the variance in people's overall assessments. When the measures were considered along with the respondent's agency and job classification, the proportion of variance increased to 21.2 percent, with people's perceptions of the ambient conditions contributing about five percentage points over and above their agency and job designation. Most likely to be dissatisfied with their agency's functional arrangement were employees in HCRS and the Post Office. Similarly, managers and professional personnel were more likely than clerical or secretarial workers to give low ratings.

People's ratings of their immediate work environment also influenced their feelings about the functional organization of the agencies in which they worked. In fact, satisfaction with the work station was the most important predictor of feelings about the agency's functional arrangement. The last part of Table 7.25 indicates that work station satisfaction added aproximately one-half to the proportion of explained variance over and above the characteristics of employees and their perceptions of distractions around them. We can conclude that people's feelings about the spatial arrangements of their agencies reflect to a large extent the way they view their immediate work environment.[18]

## Notes

1. See Chapter 3 for a discussion of the small agencies, which have been grouped together for purposes of this evaluation.

2. Postal workers were asked to skip the page with the drawings and questions about work arrangements and, therefore, the data covering federal employees are limited to only office personnel.

3. The question wording is slightly different from that found in the Harris study. In part, modifications were made to reflect two different situations — the nontraditional work settings, such as those found in the Weather Service, and the mode of question administration. The Harris study was based on face-to-face interviews, while ours used a self-administered questionnaire.

4. The methods for determining these measures are shown in the data collection form presented in Appendix C.

5. The measure assumes that all work stations within 400 square feet of the individual work stations were occupied.

6. See Figures 7.1 through 7.6 for the location of ambient measurement readings.

7. One quarter of those sampled in the national study of office workers were not very or not at all satisfied with their individual work stations. Among government employees, however, the findings were comparable to those reported here; one-third expressed some level of dissatisfaction.

8. The prior work arrangement was determined by asking respondents to indicate which of the five drawings shown in Figure 7.7 best represented the place they had previously worked. The change in office arrangement was determined by simply examining the present situation against the past and creating a new pattern variable.

9. In some cases, items from other parts of the questionnaire were shown to be conceptually and statistically related to the list of items covering work station characteristics and were included in the development of the indexes. For the inter-item correlations and reliability measures covering each index, see Appendix Table A.2.

10. Feelings about the aesthetic quality of the work station considered responses to three evaluative questions dealing with the color of walls and partitions, attractiveness, and overall aesthetic quality.

11. Furniture evaluation was based on people's responses to three questions dealing with the material used for desks, tables, and chairs, the style of the furniture, and the comfort of the chair. The reader will note that the type of chair was one of the most important objective characteristics related to work station satisfaction. Later we examine whether the type of chair assigned to an individual had any bearing on that individual's satisfaction with his or her workspace when other factors are taken into account.

12. The inter-item correlation between the two questions and the index are shown in Appendix Table A.1.

13. The two questions used to measure people's responses to their agency ambience were also used in the national study of office workers (Harris, 1978). Employees of government agencies nationally were also unhappy about their agency's appearance. More than half rated it negatively, compared to 63 percent of the Federal Building employees. Among all national office workers, 43 percent gave negative ratings to the appearance of their agency. In another national study covering all categories of workers, only 28 percent indicated their physical surroundings were unpleasant, compared to 42 percent of the Ann Arbor government workers (Quinn and Staines, 1979).

14. As in the case of attributes of the work station, items from other parts of the questionnaire that were conceptually and statistically related to items from the list of bothersome conditions were included in the development of the indexes. See Appendix Table A.3 for the items, their intercorrelations, and the coefficient of reliability for each index.

15. Several objective environmental conditions were considered vis-à-vis distractions, but few relationships were found. Among the more interesting findings: people who worked near windows were somewhat more likely to complain about the temperature being too hot in the winter and too cold in the summer than those who sat further from windows. Surprisingly, distance to the agency's entrance and the coffee pot were not related to distractions caused by the movement of people or furniture. We had expected that individuals working close to their agency's entrance and its coffee pot would be near the mainstream of pedestrian traffic and therefore would be more likely than their coworkers to complain about these distractions. Even when examining the data covering only the large agency personnel working in open-office and pool arrangements, we found no such relationships.

16. It should be pointed out that in the model considering only agency affiliation and type of job, secretaries were most likely to be bothered by noise from other agencies. However, when the ambient conditions were taken into consideration, secretaries were no more or less likely to complain than others in the building. And in the full model, the actual noise level as measured in decibels had virtually no effect on people's perceptions. The reader should note that the Beta coefficient for the agency variable in the final analysis is greater than one. This is a legitimate but rare occurrence and often reflects the presence of multicollinearity among predictors. For a full explanation of this phenomenon, see Deegan (1978).

17. In a separate analysis, we examined a number of specific ambient conditions vis-à-vis people's responses to agency ambience and found that actual levels of temperature, humidity, and light had little or no impact on ratings. In both a bivariate and multivariate context, the most important of the environmental conditions considered in predicting ambient ratings was the type of work station a person occupied. People who shared a work place in an open office gave the worst ratings.

18. One particular characteristic of the work environment employees were asked to rate was also important to their assessment of their agency's functional arrangement. People's feelings about the difficulty of access to co-workers were related to their evaluations of how well their agency functioned. Employees who indicated co-worker access was poor were most likely to give their agency arrangement low grades. This measure added four percent to the explained variance over and above work station satisfaction, employee characteristics, and their perceptions of ambient conditions.

# 8

# Worker Performance

---

## Overview

In our presentation of the conceptual model in Chapter 2, we suggested that job performance was an appropriate outcome to be examined as part of the evaluation of work environments. We also indicated that attributes of the work environment, as well as people's responses to them, could contribute to our understanding of job performance. In another chapter we noted that the efficiency and performance of workers in the new building was a recurring theme in our discussions with agency personnel. Yet, good performance was never clearly defined by the agency heads and the problems of measuring it within an office environment were readily acknowledged. We nonetheless wanted to consider it vis-à-vis the physical setting because of its importance to people in the Federal Building and its prominence as a national issue. Furthermore, the link between job performance and physical surroundings has become an accepted phenomenon. More than nine in ten office workers in the 1978 national study conducted by Louis Harris said there was a connection between job performance and personal satisfaction with the office setting.

## Environmental Conditions and Worker Performance

Within the context of the Ann Arbor Federal Building evaluation, performance was measured in three ways. First, consideration was given to how effective or productive employees believed they were in the new setting and whether their performance on the job had improved or declined as a result of the move. Second, the perceptions of the job performance of others around them was examined through direct inquiry. Finally, an index measuring the degree to which a number of ambient environmental conditions bothered or disrupted employees was created. The degree to which performance was adversely affected was inferred from the magnitude of people's complaints about the conditions around them.

*How do people's perceptions of performance*
*in different agencies vary?*

Among the questions designed to measure perceived performance on the job, one proved to be unsuccessful. More than 98 percent of the employees who were presented with the statement, "I do as much work as I reasonably can," agreed with it.[1]

We were somewhat more successful in obtaining a variety of answers to other questions dealing with performance. As seen in Table 8.1, 60 percent of the responding employees agreed with the statement, "Compared to where I worked before coming to this building, I do more work now," and 85 percent agreed that "People in my agency do as much work as they can." There were major differences among agencies in responses to the two indicators of perceived work performance. Most likely to believe they were more productive in the new building were the employees in the Weather Service (75 percent), in the military recruiters offices (73 percent), and in the Post Office (72 percent). People in IRS (54 percent), in the Social Security Administration (42 percent), and in the small agencies (56 percent) were most inclined to feel they had been more productive in their previous work environments. Employees in Social Security and the Post Office, where employees could see one another, were most likely to say their co-workers were not as productive as they might be.

*What factors are associated with people's perceptions*
*of their performance and that of others around them?*

In an exploratory effort to determine if associations exist between people's feelings about their performance in the new building and characteristics of the employees and their work environments, only one relationship was identified. Professional and technical workers tended to indicate they were less productive since the move to the new building, while the military recruiters, the postal workers, and secretarial-clerical personnel were least likely to give this response.

With respect to people's view on the performance of others, differ-

TABLE 8.1

Indicators of Perceived Work Performance, by Agency
(Percentage Distribution)

| Work Performance Indicators | All | Agency | | | | | | |
|---|---|---|---|---|---|---|---|---|
| | | Post Office | IRS | Military Recruiters | HCRS[a] | Social Security | Weather Service | Small Agencies |
| "Compared to where I worked before coming to this building, I do more work now." | | | | | | | | |
| Very true | 27 | 34 | 15 | 64 | — | 21 | 31 | 24 |
| Somewhat true | 33 | 38 | 39 | 9 | — | 21 | 44 | 32 |
| Not very true | 27 | 22 | 28 | 18 | — | 37 | 19 | 32 |
| Not at all true | 13 | 6 | 18 | 9 | — | 21 | 6 | 12 |
| Total | 100 | 100 | 100 | 100 | — | 100 | 100 | 100 |
| Number of respondents | 181 | 50 | 46 | 11 | — | 33 | 16 | 25 |
| "People in my agency do as much work as they can." | | | | | | | | |
| Very true | 38 | 37 | 35 | 46 | — | 21 | 38 | 68 |
| Somewhat true | 47 | 39 | 53 | 45 | — | 58 | 56 | 32 |
| Not very true | 12 | 18 | 10 | — | — | 18 | 6 | — |
| Not at all true | 3 | 6 | 2 | 9 | — | 3 | — | — |
| Total | 100 | 100 | 100 | 100 | — | 100 | 100 | 100 |
| Number of respondents | 185 | 51 | 49 | 11 | — | 33 | 16 | 25 |

[a] Questions were not included in questionnaires distributed to employees in Heritage Conservation and Recreation Service.

ences in opinion depended on the nature of the workspace and the degree to which people were bothered by the noise around them. Most likely to feel their co-workers were not as productive as they could be were people with a relatively small work area (60 square feet or less), people who worked within 10 feet of at least three other persons, and people who were distracted by noise. These three factors, together with the respondent's agency designation, accounted for 16 percent of the variance in the way the performance of co-workers was judged. In a multivariate context, perceived agency noise was the most important predictor, while the effects of density of the workspace on ratings of co-worker performance were negligible.[2]

### To what extent are ambient environmental conditions bothersome?

Earlier, we suggested that the degree to which people complained about the ambient conditions around them might be related to their performance on the job. That is, people's efficiency at work could be adversely affected by surrounding conditions if they were viewed as being particularly bothersome. Working under this premise, a "bothersome index" was created using the sum of the scores from the distraction items reported in the previous chapter. These items deal with people's perceptions of noise from their own and other agencies, overcompensation or undercompensation of building temperatures, glare, drafts, air quality, heat from sunlight, and the movements of people and furniture. The interrelationships between these items and the overall bothersome index are shown in Appendix Table A.4.

The average score on the bothersome index for the 239 respondents in the building was 19.5, with a standard deviation of 5.0. The worst possible situation for workers was represented by an index score of 32, while the best situation received a score of 9. The extent to which people in each agency were bothered by the composite set of conditions is shown in Table 8.2. Weather Service personnel and those in the Social Security Administration and HCRS were most bothered by ambient environmental conditions; small agency and Post Office personnel and the military recruiters were least bothered.

### What factors are associated with people's scores on the bothersome index?

The conceptual model has suggested that within the context of evaluating work environments, performance on the job is a function of the characteristics of the individual worker, his or her organization (including attributes of its physical environment), and the individual's

TABLE 8.2

Ranking of Bothersome Index Scores, by Agency[a]
(Mean Score)[b]

| Agency | Mean Score | Standard Deviation | Number of Respondents |
|---|---|---|---|
| Weather Service | 22.9 | 4.4 | 16 |
| Social Security Administration | 22.6 | 4.6 | 33 |
| HCRS | 21.7 | 4.6 | 47 |
| IRS | 20.3 | 4.9 | 49 |
| Military Recruiters | 16.7 | 4.8 | 13 |
| Post Office | 16.5 | 3.7 | 34 |
| Small Agencies | 16.1 | 3.1 | 27 |
| All | 19.5 | 5.0 | 239 |

[a] The "bothersome index" was created from individual responses to the conditions dealing with noise, temperature, glare, drafts, air quality, heat from sunlight, and the movements of people and furniture. For a review of the inter-item correlations, see Appendix Table A.4.

[b] Mean scores covering the degree to which ambient conditions are bothersome range from 9 to 32; the higher the score the more the conditions were bothersome.

feelings about the job and the work environment. Performance is also assumed to be related to overall job satisfaction. Unfortunately, measures of job satisfaction developed as part of past research (Quinn, 1977; Quinn and Staines, 1979) were not included as part of the Federal Building employees' questionnaires.[3] However, data were gathered that enabled us to examine a number of other relationships suggested by the model.

We have seen that the degree to which employees working in the Federal Building were bothered by ambient conditions varied considerably from agency to agency, with the postal workers, the military recruiters, and the people from the smaller units being the least distracted. In the other agencies, no differences were found in the index scores among people performing different jobs. However, people were bothered to various degrees depending on the exact nature of their jobs. For example, those who spent less than 75 percent of their time at their desks were considerably less bothered by ambient conditions than those who spent more time at their work stations. Similarly, workers who conversed on the telephone for less than 20 minutes each day were least bothered.

While actual ambient conditions such as noise, temperature, humidity, and light levels were unrelated to the bothersome index scores, scores did differ depending on the types of work stations people occupied and the amount of space they had. Workers in open offices and pool arrangements and those with more than 40 square feet of workspace had the worst scores on the bothersome index.

In order to see which of the several related factors were most important in explaining why scores on the bothersome index varied, a series of multivariate analyses, similar to those presented in earlier chapters, was performed (see Table 8.3). When the employee's agency and time at the desk and on the telephone were considered simultaneously, nearly one-third of the variance (29.9 percent) in the bothersome index scores was explained. The two behaviors clearly influenced the way people responded to ambient conditions, even after identifying the particular agency in which they worked. The more time people spent at their desks and on the telephone, the more likely they were to complain about ambient conditions being bothersome to their work.

The second part of Table 8.3 shows that the two objective environmental attributes accounted for only 11.5 percent of the variance in the bothersome index scores, with the quantity of space having virtually no influence, once the type of workspace the person occupied was considered. When the employee characteristics were examined along with the objective attributes, the explained variance was reduced from 29.9 percent to 24.4 percent. As noted in the footnote in Table 8.3, the reduction in the explained variance was due primarily to the loss in cases between the two analyses and the reduction in degrees of freedom resulting from the addition of two predictors. In essence, we have not improved our understanding of people's feelings about ambient conditions and the extent to which they are bothersome by knowing the type and amount of workspace they have. But the reader should note that the type of workspace, nonetheless, is important as a predictor of the bothersome index score. People in open and pool offices were clearly distracted by their surroundings, irrespective of their agency affiliation. In fact, these people were considerably distracted from their work no matter how much time they spent working at their desks or work stations.

In the final part of our explorations, we tested the proposition that people's views on the overall quality of their work environment are related to the performance on the job as measured by scores on the bothersome index. The test considered ratings for the three dimensions of the work environment discussed in Chapter 7, covering the individual work station, the agency's ambience, and its functional arrangement. Of the three, ratings of agency ambience proved to be the most powerful determinant of the bothersome index scores in both a bivariate and multivariate context. As the last part of Table 8.3 shows, agency ambience increased the explained variance to 37.4 percent.

TABLE 8.3

Bothersomeness of Ambient Conditions Predicted by Employee Characteristics, Objective Attributes and Ratings of Agency Ambience (Multiple Classification Analysis; N = 199)

| Predictors | Eta Coefficient | Beta Coefficient[a] | | | |
|---|---|---|---|---|---|
| | | Employee Characteristics Only | Objective Attributes Only | Employee Characteristics and Objective Attributes | Employee Characteristics, Objective Attributes, and Ratings of Agency Ambience |
| Employee Characteristics | | | | | |
| Agency | .53 | .43(1) | | .33(1) | .22(3) |
| Percentage of time at desk | .34 | .21(2) | | .24(3) | .20(4) |
| Minutes/day on telephone | .32 | .11(3) | | .14(5) | .16(5) |
| Objective Attributes | | | | | |
| Amount of workspace | .29 | | .12(2) | .15(4) | .15(6) |
| Work station type | .34 | | .39(1) | .29(2) | .30(2) |
| Agency Ambience Rating | .48 | | | | .44(1) |
| Percentage of variance explained | | | | | |
| (adjusted multiple $R^2$) | | 29.9[b] | 11.5 | 24.4 | 37.4 |
| (unadjusted multiple $R^2$) | | 34.5 | 15.4 | 33.2 | 46.6 |

[a] Numbers in parentheses indicate ranking of importance.

[b] The multiple classification analysis which considers only the respondent's agency and two behaviors is based on data from 213 employees. The other analyses cover only 199 employees. The loss in the multiple $R^2$ between the third analysis ($R^2 = 24.4$) and the first ($R^2 = 29.9$) reflects this difference in the number of respondents. Had the respondents been identical in both analyses, the difference in variance explained would not have been as great. The $R^2$s would never be identical since the second analysis which adds two predictors has lost several degrees of freedom, which is reflected in the adjusted $R^2$. If data covering the same respondents had been used, the analyses would result in identical values for the unadjusted $R^2$. In fact, it can be seen that the unadjusted $R^2$ between the first and third analyses are considerably closer in value.

Based on these analyses, we can conclude that relationships suggested by the conceptual model do exist to varying degrees and that if the extent to which people are bothered by aspects of their physical surroundings is an appropriate indicator of performance, then objective environmental conditions and people's feelings about them can affect the quality and quantity of their work.

## Notes

1. The question was answered by only 189 employees. In addition to 3 non-responses, 47 HCRS employees were not given questionnaires containing this and other questions dealing with job performance or other non-environmental issues. In retrospect, it seems obvious that most people would not openly admit to being less than fully productive in their work. Thus, we should not have been too surprised by the distribution of responses to this question.

2. The analysis of data covering perceptions of co-worker performance was based on all agencies except the HCRS and the Post Office. Data were not available for the former, while the postal workers were excluded because of the idiosyncratic nature of their work-spaces and the work they do.

3. As noted earlier, we were unsuccessful in persuading individuals whose cooperation was essential to the execution of the study that it was important to ask questions related to job satisfaction and other non-environmental issues.

# 9

# Conclusions and Recommendations

---

In the preceding chapters, we have presented detailed findings from an evaluation of the Ann Arbor Federal Building. These findings deal with relationships between the building and its surroundings, transportation and parking, the architectural quality of the building, the work environment, and worker performance.

For the most part, the findings are based on analyses of quantitative data covering the federal workers, their attitudes and behaviors, and the characteristics of the environments in which they work. The analyses have been guided by a conceptual model suggesting the manner in which workers, their organizations, and the physical surroundings interact.

In this final chapter, we summarize the key findings and use them in drawing conclusions about the Ann Arbor Federal Building and the degree to which it is successful. We do so by considering the findings in light of the specific objectives the GSA representatives and their architects had hoped to accomplish through their programming and design efforts. Building on these findings, we then outline a number of recommendations for governmental officials, architects, and space planners concerned with the quality of work environments. Finally, we suggest several avenues for evaluation research on built environments that are worthy of future consideration by environmental design researchers and policy makers. Before doing so, however, we review the environmental changes in the Federal Building that have taken place since the evaluation was completed.

## Environmental Change

Evaluators of any built environment should always expect that changes will occur between the time of the initial reconnaissance and the completion of systematic data collection. Similarly, environmental changes are likely to take place after the data are analyzed but before the findings are publicized. Changes in the Ann Arbor Federal Building have occurred during both periods. Some of the changes were alluded to in Chapter 2. At a general level, we found that over the course of the three months of data collection, the location of people and furniture had been rearranged in several agencies. Two weeks prior to the distribution of questionnaires, the heating and ventilating system was modified and the ambient conditions were not fully stabilized by the time questions about heating and ventilation were being answered. Nor did employee responses in November reflect actual ambient conditions measured three months later.

After the data-gathering phase of this study, other major changes occurred in the building that are not reflected in the analyses presented here. These changes involve the departure of agencies from the building, improved parking, a new signboard system, the introduction of background music into the work environment, and staff turnover.

*Agency departures.* In early 1980, GSA announced that the Federal District Court which was to occupy the building would move into the first and fourth floors and that an architectural firm had been commissioned to prepare plans for the necessary renovations. Subsequently, the military recruiters on the first floor and two of the small agencies on the fourth floor vacated the building, while a search for new and comparable office space for the Internal Revenue Service was begun by GSA. In December 1980, the National Power Plant Team, a small federal agency scheduled to move from the region in April 1981, was housed temporarily in the fourth-floor space vacated by the Soil Conservation Service three months earlier. A visit to that agency shortly after the first of the year revealed an attractive work environment, which team members said they liked despite its temporary nature.

The prospective move of IRS outside of Ann Arbor's central business district caused considerable dismay among local officials and merchants, who viewed the relocation as contrary to the city's revitalization efforts. Several attempts were made by city officials to assist GSA in finding sufficient space for IRS in the central area, all to no avail. This large agency was scheduled to move from the building and from downtown Ann Arbor in the late spring of 1981.

*Parking.* In response to a congressional inquiry concerning parking congestion at the Ann Arbor Federal Building, GSA made modifications to the rear parking lot to provide 15 additional spaces for public use. Our subsequent observations of the situation revealed that traffic congestion caused by queuing at the entrance of the short-term lot along Fifth Avenue had been substantially reduced.

*Signboards.* It has been noted that federal workers gave low marks to the quality of signs in the building. Indeed, during the data collection period, signs of various sizes and shapes and with different types of lettering were found throughout the building. Unfortunately, these temporary signs were in use for more than two years before a more attractive, legible, and uniform signage system was installed in the fall of 1980.

*Music.* Shortly after the completion of the environmental data collection, Muzak was introduced into the building's intercom system. The installation was intended to improve the quality of the work environment by introducing background music whose volume could be controlled within each agency. We have no way of knowing at this writing where and how frequently the system has been used or how federal workers or the public feel about it.

*Staff turnover.* Finally, there have been varying amounts of staff turnover in the agencies housed in the building. In some, the staff has remained unchanged, while in others considerable turnover has been reported. We do not know the exact nature of personnel change within each agency. Nor do we know if and how individual workspaces have been altered to accommodate these changes or to improve the quality of the work environment.

## Conclusions and an Overview of Findings

We can conclude that the Ann Arbor Federal Building is successful in one major respect — it has become an integral part of downtown Ann Arbor and has contributed to the attractiveness and economic vitality of the area. It is readily identifiable and used by the general public. Most community residents consider it attractive, worthy of its many awards for design excellence, and conveniently located. For the most part, the federal employees there also like the location and take advantage of downtown and campus shopping, restaurants, and other central area facilities.

On the other hand, the building has not lived up to its expectations of providing a high quality work environment for all of its occupants.

The majority of employees rated the workspaces of their agencies as only fair or poor, while a substantial number were dissatisfied with the particular place they occupied.

These conclusions are based on a review of the designers' initial objectives, outlined in Chapter 2, and of the specific findings pertaining to them.

> The building should be an integral part of downtown Ann Arbor. It should be in visual harmony with the character of the downtown setting and a catalyst for new downtown development.

More than eight in ten Ann Arbor residents knew the location of the Federal Building and three-quarters had visited it at one time or another. Most people said it fit into its surroundings, and they particularly liked its plaza and the setback along Liberty Street. While there are no data to support this contention, it appears that the building has served as a catalyst for new downtown development. Since the facility was first announced, renovations of older structures have been prevalent within a two-block radius, and a new commercial building to the east of the Ann Arbor Federal Building was built about three years ago.

> Interaction between building occupants and patrons and the downtown community should be fostered. It should be functionally a part of downtown Ann Arbor and should be used extensively by community residents. It should be a stopping point for pedestrians who travel along Liberty Street between downtown Ann Arbor and the University of Michigan campus.

Significant numbers of users of the building worked or conducted personal business downtown, and many were U-M students. A third of the users walked to the building and a comparable proportion made use of the plaza. The public was most inclined to visit the Post Office and, to a lesser extent, the Social Security Administration, the IRS office, and the military recruiters. Other agencies were rarely visited. Neither the public nor building employees extensively used the coffee shop or its lounge facilities.

> The building should exemplify good architectural design without being a dominating or imposing structure.

Three out of every four members of the general public thought the building was attractive and worthy of its architectural honors. A somewhat smaller proportion liked the interior. Building occupants, on the other hand, were likely to give it low marks both on architectural quality and as a place to work.

The work spaces within the building should allow for flexibility and change, both within agencies and throughout the building as a whole. Flexibility should be accomplished without hindering the performance of workers. The structure should be designed to eventually house a federal district court facility.

Opportunities for changing the workspace were limited for the smaller agencies consisting of conventional offices, and little change was noted. Rearranging of furniture, however, did take place with ease and regularity in the open offices characteristic of the larger agencies. Yet the flexibility inherent in open offices was not without costs. One-fourth of the workers in open offices and a third of those in pool offices indicated the movement of furniture around them was bothersome and hindered their job performance.

The building should be designed so as to create a sense of community among the people who work there.

Two design features were intended to promote this feeling among federal employees: the lounge in the second-floor lobby and the open lightwells between adjacent agency spaces. We have already noted that relatively little use was made of the lounge. It was never furnished nor decorated in the manner specified by the architects, and the hopes of creating an attractive and inviting space were never realized.

The open lightwells were a source of annoyance rather than an attraction. Many people at work stations below a lightwell and another agency were vociferous in their complaints about noise and the lack of privacy. This situation was hardly conducive to a sense of community.

We do not know how workers actually felt about employees from other agencies nor, for that matter, about their co-workers. The data do show extensive employee interaction within agencies, and access to co-workers was rated favorably. However, few employees indicated they visited other organizations in the building. We suggest that in a building that serves different functions and contains diverse groups of individuals and organizations, fostering a sense of community is an unrealistic objective.

Employees should take pride and find satisfaction in their work environment.

The federal employees were of mixed minds in their assessments of their work environments. While most expressed some level of satisfaction with the workspace available to them and the overall ambience of their agencies, many were dissatisfied with their physical surroundings. A third were dissatisfied with their immediate workspace and a quarter gave poor ratings to both the appearance and the spatial arrangement of their agency. To a large extent, dissatisfaction was associated with little privacy, poor views, temperature variability, and distractions caused by noise from other agencies. Dissatisfaction was most prevalent among workers in open and pool offices.

> Opportunities should be provided for employees to store personal belongings and to personalize their workspaces according to individual tastes and interests. Work areas should be functional, efficient, and conducive to agency work requirements.

Many employees viewed storage areas as barely adequate, and people in open and pool offices felt that surface areas for hanging things was insufficient. Nonetheless, about half of the work stations were personalized with plants, pictures, or desk paraphernalia. Whether more storage and larger surface areas would result in more personalization is subject to speculation.

> The building should be designed as an energy-efficient structure. It should be oriented so as to take advantage of the natural lighting on the north and to minimize heat gain on the east and west.

We have no way of knowing the extent to which the building is energy efficient. Despite federal guidelines requiring temperatures below 68 degrees in winter, an average building temperature of 74 degrees was recorded during the evaluation. Attempts to minimize heat gain by eliminating windows on the south, east, and west resulted in workers complaining about lack of views in the south part of the large, open-office areas.

> Materials should be selected so as to inhibit vandalism and reduce maintenance costs.

No attempts to assess this objective were made as a part of this evaluation. We did observe during the year-and-a-half study, however, that the exterior and interior of the building were well maintained and there were no apparent signs of vandalism.

### Policy Recommendations

Based on the results of the evaluation, a number of recommendations can be made for alleviating some of the problems that have been identified. The findings point to possible guidelines for programming, designing, and managing work environments and federal office buildings in other settings.

As we noted, many of the problems that existed have been resolved in recent months. Others have not. For example, traffic congestion along Fifth Avenue and public parking have both improved as a result of converting employee spaces behind the building for public use. The situation might even be better if signboards indicating the availability of additional parking were to be posted in both short-term lots.

New signboards have improved the visual quality of the building's interior and, we suspect, have made it easier for people to locate specific places.[1] If not, diagrams of floor plans of the building would be a useful addition to the directory in the main lobby.

Problems associated with the mechanical system, reflected in responses of workers in November 1979, have since been rectified. It is not known, however, whether the improvements are uniformly recognized or if people from different agencies experience the same ambient conditions. Recently, we have heard complaints in some agencies about excessively warm temperatures.

At a more general level, it is clear that knowing and understanding the organizations expected to occupy a building are necessary prerequisites for developing or at least finalizing a design concept. While the vertical flow of space between floors may be appropriate within one organization or for organizations that are functionally compatible, the concept appears unworkable in a federal building accommodating many diverse agencies.

In open offices, the problem of visual and conversational privacy, space, views, and the location of electrical and communication outlets should be recognized as critical to the workers' environmental satisfaction. These problems became particularly acute when flexible furniture systems specified by the architects and promised to the workers were not used. Old furnishings and improper moveable partitions clearly contributed to worker dissatisfaction.

In the planning of the building, the architects showed great sensitivity to the needs of the occupants. Agency personnel were surveyed during the design stage to determine job requirements and space preferences. This information was then used to plan the furniture arrangements and determine the most appropriate furniture system to supplement the open-office concept. We do not know the extent to which environmental problems would have existed had the specified furniture system been installed at the time federal workers moved into the building. Nor do we know how people would respond to the work environment if the system were installed today. We suspect some problems would still prevail. For the future, we suggest that attempts to seek user input continue and that the planning of interior spaces reflect the information provided. We suggest that this process be performed with great care, particularly if there is any possibility that the expectations created among the users will not be met.

The successful integration of the Federal Building into downtown Ann Arbor, makes it appear that GSA's current guidelines regarding the central location of federal facilities are appropriate. Certainly the location of government buildings in places that are readily accessible to the public by car, via public transportation, or on foot is highly desirable. At the same time, buildings that are kept in scale with the surroundings are recognized as attractive additions to the urban scene. The provision of usable outdoor open space can be an important amenity within central business districts.

Consideration should be given to the parking needs of both the public and building users in choosing a downtown location. While the building conveniently serves those who ride a bus or walk and drivers who visit the building for only short periods of time, parking can be a problem for long-term visitors. The agency personnel could also help visitors by informing them in advance, if possible, about how long their meetings would last. Nearby parking with long-term meters is essential.

We recognize that free parking for federal employees in a downtown location is prohibitive. We are also sympathetic to high parking costs for employees who must drive. Some former drivers have adjusted to the situation by carpooling, walking, or using public transportation. The frustrations of parking are exacerbated by the fact that most employees had free parking available to them at their previous places of work. Under such circumstances, it seems important that workers be informed prior to their move about any disadvantages they might encounter, such as parking problems, as well as advantages, such as the attractions of a downtown location.

## Research Recommendations

In addition to specific findings that may be applicable to other settings and the guidelines that can be used in building planning and management, we are able to offer several suggestions for improving the process of evaluating built environments. Some are derived from our past experiences, while others stem from limitations identified as part of this work.

As we noted in Chapter 2, there was a lapse in time between the gathering of evaluative data and the collection of objective environmental measures. As a result, the environmental conditions in our data set were not always reflective of the questionnaire responses we had obtained ten weeks earlier. Furthermore, both data gathering efforts occurred only during the winter and not during other seasons when ambient conditions and people's responses to them might have been quite different. In part, the time lapse between the two data collection periods reflected a shortage of trained personnel to carry out the necessary work. It also resulted from delays in obtaining the technical instruments ordered in connection with the project. Once the instruments were received, the research team had to be trained in their use, resulting in further delays. We do not know how these problems could have been avoided; we suspect that better planning for data requirements and the equipment and manpower used in obtaining the data would have eliminated some of them. Nonetheless, it is important to consider such matters when contemplating future evaluation studies. At the same time, serious efforts should be made concurrently to obtain the various kinds of data that are intended to be examined in relation to each other.

We noted a limitation in collecting objective data on ambient conditions during a single season. Additionally, no more than two readings for each condition were obtained at one time and usually within a one-week period. Ideally, we would want to gather ambient environmental data throughout the year and within specified time periods in order to reflect variability in seasons, in the time of day, and in outside conditions. Clearly, developmental work is needed to improve procedures for gathering data about building conditions.

As part of that work, special attention should be given to developing techniques for systematically and quickly measuring light and glare conditions. Our attempts to measure lighting were moderately successful. Nonetheless, the measurements took more time than we had expected. Our efforts to measure glare were fraught with problems and we ultimately resorted to a measure of glare condition based largely on desk orientation and window location.

In the area of job performance, the need for better measures is widely recognized. In the context of our study, we relied primarily on the workers' perceptions of their own productivity and that of others around them. We also suggested that an individual's score on the "bothersome index" was an indicator of job performance; the index was based on people's perceptions of the environmental conditions around them.

No attempts were made to use objective measures of job performance developed by others. Nor were we able to develop our own objective indicators, largely because of restrictions placed on us by agency personnel whose cooperation was essential in executing the evaluation. We were also asked not to examine job satisfaction or some other work-related attitudes. Our model indicates that such measures are essential to understanding the dynamics of the work environment. In future evaluations dealing with work environments, we suggest that stronger efforts be made to include measures dealing with job satisfaction and the organizational context within which jobs are performed. A statement of the rationale for asking job-related questions as part of a building evaluation should also be prepared.

We are not certain whether all our objective measures of the work environment were appropriate. To a large extent, we view our evaluation as developmental, and perhaps our objective data collection was excessive in scope. At the same time, additional data characterizing the work environment might have been gathered. For example, it might be appropriate to know where people are located vis-à-vis the specific equipment they use and co-workers they meet regularly. Or it may be useful to measure air flow or record the color of work surfaces. Clearly, more attention should be given to developing a battery of appropriate measures describing work environments and their specific attributes.

We know from our analysis that responses to a number of questions differed depending on the job classification and sex of workers. These two items represent person characteristics shown in the conceptual model in Chapter 2. Undoubtedly, other characteristics of individuals would influence their responses to environmental conditions. Within the context of research on person-environment relations, including evaluation studies, efforts should be made to identify these other characteristics that may act as a mediating influence on people's responses.

We noted in Chapter 7 that some employees had difficulty in interpreting the drawings used in the questionnaire. The reader is reminded that these drawings were used in the national study of office workers and were selected for comparative purposes. We suspect that the drawings would present difficulties for respondents in other settings as well. Yet drawings within questionnaires can be useful in conveying ideas, particularly those dealing with the physical environment. In light of this, we believe that more basic research is needed on the appropriateness of graphics in eliciting responses of people occupying work and other environments.

Earlier, we mentioned that our efforts to disseminate preliminary findings met with little interest. We are not certain whether this reflects the findings themselves, the mode of presentation, or the environmental setting within which we were working. Our experience suggests that care should be taken in planning the content and process of the feedback of evaluative data and that alternative dissemination strategies be developed and tested in a variety of settings.

Finally, we suggest that efforts should be made to use the results of evaluations in developing patterns of environmental attributes that in turn can be used by designers in their planning activities.[2] Rarely are designers able to work with discrete measures of environmental conditions and people's responses to them in designing buildings or other places. Architects, for example, cannot assume that placing a worker in proximity to a window will guarantee occupant satisfaction with the work station. That decision must be balanced with other design considerations, some of which may conflict with the original decision. The placement of windows in a building, while offering people a view, tends to increase energy use and creates high perimeter ratios. Windows may also be a source of distraction for certain work-related tasks.

In our Federal Building evaluation, we have employed several discrete environmental measures and have shown how employees responded to them. We have also examined several measures simultaneously in order to see their combined effects on people's attitudes and behaviors. In a more exploratory effort, we predefined a number of work settings or patterns within the building and examined occupants' responses to a variety of environmental conditions within each. In essence, we have attempted to develop an empirically based definition of patterns of work environments in one particular building.

Eight patterns or environmental zones were characterized. These are shown in Figure 9.1. Two basic factors guided our zonal characterization: an interest in reducing environmental data to a form understand-

FIGURE 9.1

Schematic Environmental Zones*

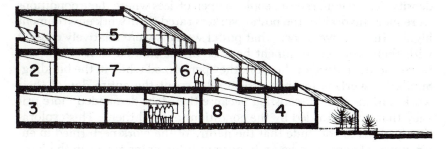

*Diagram is not a true representation of the building section.

able to architects, and a recognition that in past studies office workers have responded to a limited number of interrelated variables, such as the presence or lack of views, lighting levels, and work station privacy.

Two environmental measures were used to define the zones: the type of work setting and the degree to which the work stations in those settings were connected to the exterior by sunlight or views. We defined three types of work settings — industrial, conventional, and open offices — and three kinds of connections to the exterior environment — light and view, light and no view, and closed. The precise definition of these conditions is shown in Appendix D, while Table 9.1 summarizes the characteristics of the zones.

In order to determine if user responses were related to their zonal locations within the building, a number of bivariate analyses were considered. A sampling of the results is shown in Figure 9.2. Average evaluation scores for two components of the work environment — agency ambience and its functional arrangement — are presented for occupants in each of the eight environmental zones. The figure also presents average scores covering the degree to which employees were bothered by noise from within their own agency. In each instance, higher levels of employee satisfaction were found in conventional office settings than in open offices. Furthermore, workers in offices with vertical connections to adjacent agencies (Zones 6 and 8) were more likely than workers without such a connection or workers without natural light or views to give negative responses. These findings are in line with the data presented in Chapter 7; conventional or closed-office settings seem to be more conducive to worker satisfaction than open offices.

The findings also show that a setting having limited attributes conducive to a satisfactory work environment can be evaluated relatively favorably by its workers. The industrial environment occupied by the postal carriers lacked a view to the outside, had no natural light, was devoid of acoustical treatment, and had high levels of noise and worker density. Yet, compared to people in open offices where these conditions were more favorable, the postal workers rated their work environment highly. In part, we suspect that postal workers were relatively content with their work environment because of the limited time they spent there and the nature of the work they performed while in the building. Similarly, workers in open offices who spent the entire day at their work stations were dissatisfied because their jobs demanded more privacy than was afforded by the conditions around them. This exploration leads us to conclude that the nature of the work performed in environmental zones can be an important intervening factor in the satisfaction people derive from their physical settings. In fact, the role of an

TABLE 9.1

Description of Environmental Zones

| Zone | Work Setting | | | Outside Connection | | | Description | Number of Work Stations |
|------|-------------|--|--|-------------------|--|--|-------------|------------------------|
| | Industrial | Conventional | Open Office | Natural Light and View | Natural Light Only | None | | |
| 1 | | ● | | ● | | | Small private office with outside connection | 20 |
| 2 | | ● | | | | ● | Small private office without outside connection | 28 |
| 3 | ● | | | | | ● | Postal mail sorting | 44 |
| 4 | | | ● | | ● | | Open office within 10 feet of lightwell and no agency above | 22 |
| 5 | | | ● | ● | | | Open office with direct view of window and no agency below | 28 |
| 6 | | | ● | ● | | | Open office with direct view of window and agency below window | 18 |
| 7 | | | ● | | | ● | Open office with no direct view of window and more than 10 feet from lightwell | 74 |
| 8 | | | ● | | ● | | Open office within 10 feet of lightwell and agency above | 26 |

FIGURE 9.2

Relationships between Occupants' Evaluations
and Environmental Zones

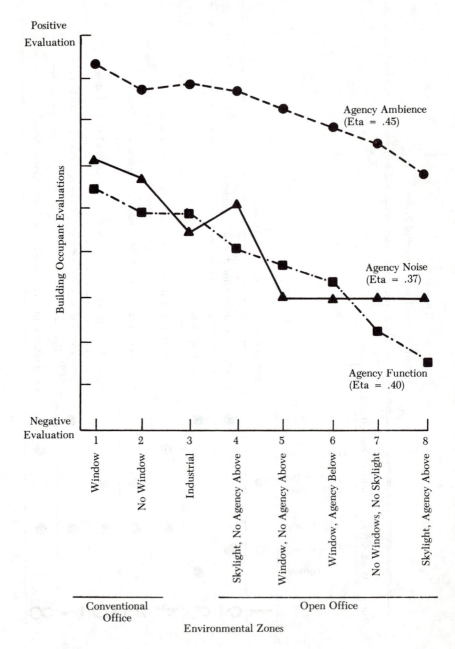

individual in terms of his or her work in explaining environmental assessments has been suggested by the model presented in Chapter 2. Furthermore, the model suggested that an individual's organization, including the tasks taking place within it, will interact with environmental conditions to influence performance. The testing of this proposition should be an important goal in future evaluations of work
environments.

This chapter has presented the results of a systematic evaluation of a
built environment. We have examined the degree to which specific
purposes and objectives of the environment have been met. Each objective gleaned from the building's sponsors and designers was considered in light of findings from an analysis of data covering the building and its users. A number of recommendations have been made on
the basis of the evaluation. These include the alleviation of problems
identified as part of the work, general guidelines for programming,
designing, and managing work environments, and federal office buildings in other settings and directions for future research.

## Notes

1. An evaluation to test this hypothesis might be in order.
2. Christopher Alexander and his associates have referred to the use of "patterns"
when describing packages of spatial concepts that can be used in researching or creating
design solutions (Alexander et al., 1977). The patterns are composed of many physical
attributes, which together define an environmental setting. That setting is likely to yield
particular behavior responses by its users. For instance, a small work group is a pattern
used to describe an environment with less than half a dozen people and a set of physical
attributes. The pattern is reported to be functionally optimal and most satisfying to the
occupants. Alexander based his patterns on a combination of empirical research reported
by others and anthropological investigations.

# Appendix A

## Supplementary Tables and Figures

---

APPENDIX TABLE A.1

Building and Agency Evaluation Indexes
(Product-Moment Correlations)

| Evaluation Indexes | Index | (A) | (B) | (C) | (D) | Reliability Coefficient[a] |
|---|---|---|---|---|---|---|
| Building Architectural Quality | | | | | | |
| (A) Attractive-unattractive (8a)[b] | .67 | | | | | |
| (B) Good-poor design (8i) | .72 | .55 | | | | |
| (C) Stimulating-unstimulating spaces (8k) | .70 | .44 | .50 | | | |
| (D) Pleasant-unpleasant (8f) | .70 | .55 | .44 | .41 | | |
| (E) Architectural quality (8d) | .76 | .43 | .59 | .39 | .44 | .65 |
| Building Upkeep | | | | | | |
| (A) Interior of building (8b) | .89 | | | | | |
| (B) Exterior of building (8c) | .86 | .51 | | | | .69 |
| Agency Ambience | | | | | | |
| (A) Space appearance (12) | .90 | | | | | |
| (B) Pleasant physical surroundings (25j) | .89 | .62 | | | | .75 |

[a] For a discussion of the coefficient of reliability, see Nunnally (1967).

[b] Numbers in parentheses refer to questionnaire items.

## APPENDIX TABLE A.2

### Work Station Characteristic Evaluation Indexes
### (Product-Moment Correlations)

| Evaluation Indexes | Index | (A) | (B) | Reliability Coefficient[a] |
|---|---|---|---|---|
| **Lighting** | | | | |
| (A) Lighting for work (23c)[b] | .92 | | | |
| (B) Location of ceiling (23d) | .88 | .85 | | |
| (C) Ceiling light glare (16k) | .77 | .53 | .44 | .82 |
| **Space** | | | | |
| (A) Space available (23a) | .87 | | | |
| (B) Surface area for work (23q) | .87 | .68 | | |
| (C) Space for storage (23f) | .81 | .54 | .56 | .81 |
| **Aesthetics** | | | | |
| (A) Attractiveness (23g) | .88 | | | |
| (B) Overall aesthetic quality (23s) | .83 | .67 | | |
| (C) Color of walls and partitions (23e) | .82 | .57 | .48 | .80 |
| **Electrical Outlets** | | | | |
| (A) Number (23n) | .95 | | | |
| (B) Location (23o) | .94 | .80 | | .89 |
| **Conversational Privacy** | | | | |
| (A) Conversational privacy (23h) | .78 | | | |
| (B) Hear co-worker discussions (25c) | .84 | .46 | | |
| (C) Hear telephone conversations (25g) | .83 | .44 | .68 | .75 |
| **Furniture** | | | | |
| (A) Materials for desks, tables and chairs (23b) | .74 | | | |
| (B) Furniture style (23m) | .83 | .47 | | |
| (C) Comfort of chair (23r) | .80 | .34 | .51 | .70 |

[a] For a discussion of the coefficient of reliability, see Nunnally (1967).

[b] Numbers in parentheses refer to questionnaire items.

APPENDIX TABLE A.3

Individual Bothersome Indexes
(Product-Moment Correlations)

| Bothersome Indexes | Index | (A) | (B) | Reliability Coefficient[a] |
|---|---|---|---|---|
| Own Agency Noise | | | | |
| (A) Telephone bothersome (16a)[b] | .79 | | | |
| (B) Equipment bothersome (16c) | .79 | .47 | | |
| (C) Talk bothersome (16e) | .80 | .48 | .44 | .72 |
| Other Agency Noise | | | | |
| (A) Telephones bothersome (16b) | .87 | | | |
| (B) Equipment bothersome (16d) | .92 | .71 | | |
| (C) Talk bothersome (16f) | .89 | .63 | .78 | .88 |
| Temperature Overcompensation | | | | |
| (A) Hot in winter (16o) | .18 | | | |
| (B) Cold in summer (16m) | .26 | .54 | | .60 |
| Temperature Undercompensation | | | | |
| (A) Hot in summer (16l) | .26 | | | |
| (B) Cold in winter (16n) | .20 | .45 | | .70 |
| Distraction | | | | |
| (A) People walking around (16s) | .84 | | | |
| (B) Furniture rearrangement (16t) | .84 | .43 | | .59 |
| Air Quality | | | | |
| (A) Stuffy air (16r) | .86 | | | |
| (B) Ventilation and air circulation (23t) | .87 | .58 | | |
| (C) Air quality (23v) | .86 | .60 | .77 | .82 |
| Glare | | | | |
| (A) Glare from natural light (16j) | .77 | | | |
| (B) Glare from ceiling lights (16k) | .88 | .33 | | .50 |

[a] For a discussion of the coefficient of reliability, see Nunnally (1967).

[b] Numbers in parentheses refer to questionnaire items.

## APPENDIX TABLE A.4

### Overall Bothersome Index
(Product-Moment Correlation)

| Overall Bothersome Index | Index | Individual Items and Bothersome Indexes | | | | | | | | Reliability Coefficient[a] |
|---|---|---|---|---|---|---|---|---|---|---|
| | | A | B | C | D | E | F | G | H | |
| (A) Own agency noise | .55 | | | | | | | | | .54 |
| (B) Other agency noise | .52 | .14 | | | | | | | | |
| (C) Temperature overcompensation | .58 | .18 | -.03 | | | | | | | |
| (D) Temperature undercompensation | .48 | .07 | .04 | .26 | | | | | | |
| (E) Distractions | .60 | .54 | .24 | .22 | .10 | | | | | |
| (F) Glare | .55 | .19 | .20 | .15 | .16 | .32 | | | | |
| (G) Stuffy air | .67 | .25 | .11 | .35 | .38 | .33 | .25 | | | |
| (H) Drafts | .60 | .27 | -.04 | .44 | .21 | .15 | .33 | .28 | | |
| (I) Heat from the sun | .55 | .13 | .00 | .37 | .15 | .30 | .32 | .23 | .27 | |

[a] For a discussion of the coefficient of reliability, see Nunnally (1967).

## APPENDIX FIGURE A.1

### Average Ratings of Personal Work Station Characteristics
### (For Post Office and Building as a Whole)

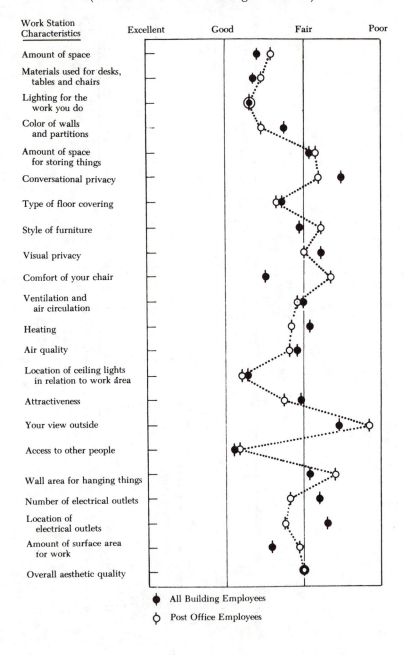

APPENDIX FIGURE A.2

Average Ratings of Personal Work Station Characteristics
(For IRS and Building as a Whole)

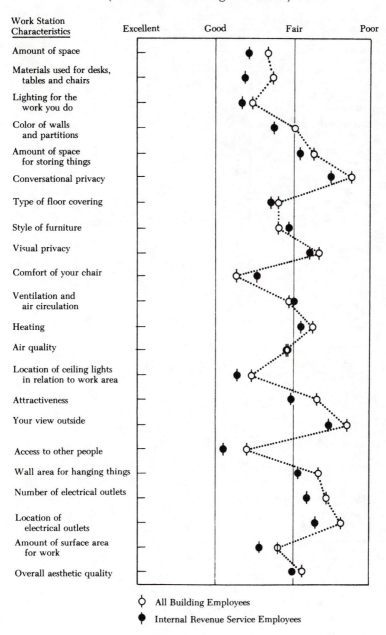

○  All Building Employees

●  Internal Revenue Service Employees

APPENDIX FIGURE A.3

Average Ratings of Personal Work Station Characteristics
(For Military Recruiters and Building as a Whole)

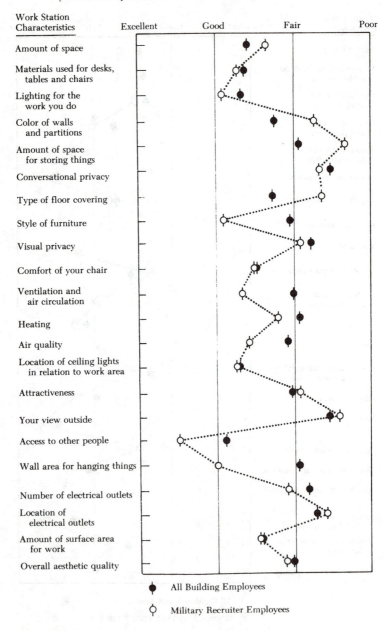

● All Building Employees

○ Military Recruiter Employees

# APPENDIX FIGURE A.4

## Average Ratings of Personal Work Station Characteristics
### (For HCRS and Building as a Whole)

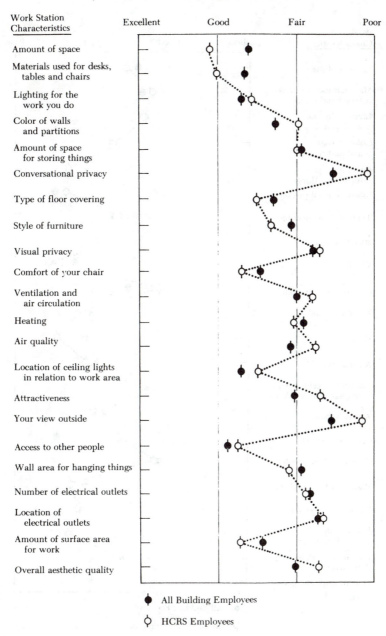

| Work Station Characteristics | Excellent | Good | Fair | Poor |
|---|---|---|---|---|
| Amount of space | | | | |
| Materials used for desks, tables and chairs | | | | |
| Lighting for the work you do | | | | |
| Color of walls and partitions | | | | |
| Amount of space for storing things | | | | |
| Conversational privacy | | | | |
| Type of floor covering | | | | |
| Style of furniture | | | | |
| Visual privacy | | | | |
| Comfort of your chair | | | | |
| Ventilation and air circulation | | | | |
| Heating | | | | |
| Air quality | | | | |
| Location of ceiling lights in relation to work area | | | | |
| Attractiveness | | | | |
| Your view outside | | | | |
| Access to other people | | | | |
| Wall area for hanging things | | | | |
| Number of electrical outlets | | | | |
| Location of electrical outlets | | | | |
| Amount of surface area for work | | | | |
| Overall aesthetic quality | | | | |

●  All Building Employees

○  HCRS Employees

## APPENDIX FIGURE A.5

### Average Ratings of Personal Work Station Characteristics
### (For Social Security and Building as a Whole)

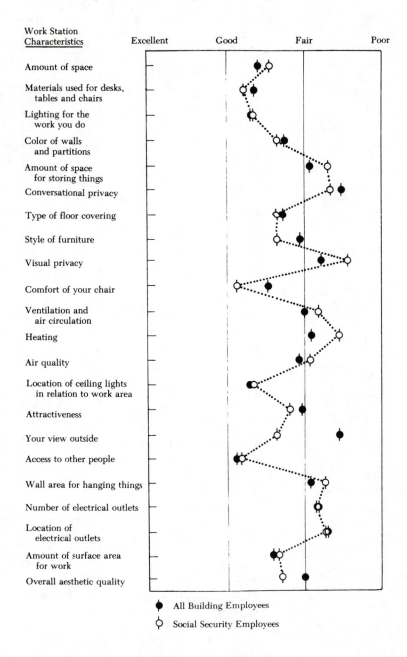

● All Building Employees

○ Social Security Employees

# APPENDIX FIGURE A.6

## Average Ratings of Personal Work Station Characteristics
## (For Weather Bureau and Building as a Whole)

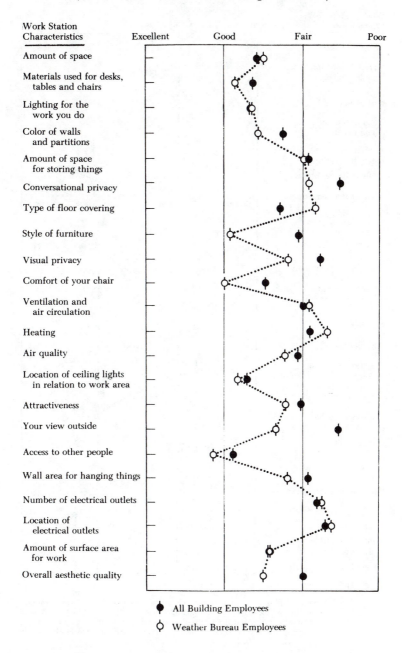

● All Building Employees

○ Weather Bureau Employees

APPENDIX FIGURE A.7

Average Ratings of Personal Work Station Characteristics
(For the Small Agencies and Building as a Whole)

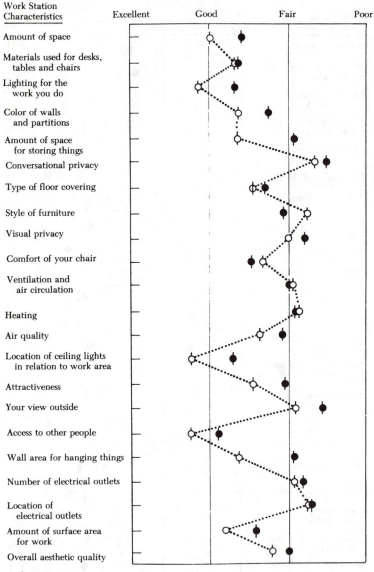

● All Building Employees

○ Small Agency Employees: Department of Defense-Army Recruiting Area Commander;
Defense Logistics Agency; Defense Investigative Service and Army Surgeon General;
Soil Conservation Service; District Court-Probation Department; Department of
Labor – Wage and Hourly Division; and the Federal Bureau of Investigation.

# Appendix B

## Questionnaires

## College of Architecture & Urban Planning

The University of Michigan  Ann Arbor, Michigan 48105  313/764-1340

Architectural Research Laboratory

November, 1979

Dear Federal Building Occupants:

The University of Michigan's College of Architecture and Urban Planning is in the process of conducting an evaluation of the Ann Arbor Federal Building. This project, sponsored by the National Bureau of Standards of the Department of Commerce, is an attempt to learn how well the buildings meets the needs of its occupants. The information we obtain can be useful to the federal government in the design and planning of new federal buildings elsewhere. At the same time, we expect to produce information that can be used in making modifications to the Ann Arbor building so as to improve the quality of the work environment for present and future building occupants.

As part of this study, we will distribute questionnaires to each of you during the next week or so. At the same time, members of the research team will visit each agency to collect information about specific environmental conditions. In order to make this a successful project, your cooperation is essential. We would like you to fill out the questionnaire as carefully as possible on the same day they are distributed. If there is any question that you are unable to answer or don't want to answer, just skip it and go on to the next one. Naturally, your responses to the questions will remain anonymous. No one will see the questionnaires except members of the research team and the results will be tabulated in statistical form covering everyone in the building.

While we do not want to identify individuals by name, we do need to know where people work in the building so that the necessary measures of environmental conditions (lighting, heating levels, etc.) can be taken and compared with peoples' responses. Therefore, the questionnaire you will receive will have an identification number which corresponds to your work area.

In order to guarantee that no one besides members of the research team will see your completed questionnaires, a collection box will be placed at a designated location within your agency for you to deposit your questionnaires after you have completed them.

We thank you in advance for your cooperation.

Very truly yours,

Robert W. Marans, Professor
College of Architecture and Urban Planning
Director, Federal Building Evaluation Project

RWM*bz

---

SPACE ID NO. _____

INTERVIEW NO. _____

**College of** Architecture & Urban Planning

The University of Michigan  Ann Arbor, Michigan 48105
Architectural Research Laboratory

### FEDERAL BUILDING OCCUPANTS QUESTIONNAIRE

27 November 1979

Dear Federal Building Occupants:

This is the questionnaire we told you about several days ago. As you may recall, it is designed to help us in our evaluation of the Federal Building conducted under National Bureau of Standards sponsorship. Please fill it out as completely as possible and return it to your agency's collection station by the end of the week.

If there is any question that you are unable to answer or don't want to answer, just skip it and go on to the next one. Thank you for your cooperation.

Sincerely,

Robert W. Marans, Professor
College of Architecture and Urban Planning
Director, Federal Building Evaluation Project

1. How long have you worked for your agency?

   ☐ LESS THAN 1 YEAR
   ☐ 1 - 2 YEARS
   ☐ MORE THAN 2 YEARS, BUT LESS THAN 5 YEARS
   ☐ 5 - 10 YEARS
   ☐ MORE THAN 10 YEARS

2. How long have you worked in the Ann Arbor Federal Building?

   ☐ LESS THAN 3 MONTHS
   ☐ 3 TO 6 MONTHS
   ☐ MORE THAN 6 MONTHS, LESS THAN 1 YEAR
   ☐ 1 - 2 YEARS
   ☐ MORE THAN 2 YEARS

**2**

3. How do you usually get to and from work? (CHOOSE ONE)
[ ] OWN CAR
[ ] AGENCY CAR
[ ] SHARE RIDE OR CAR POOL
[ ] BUS
[ ] WALK
[ ] BICYCLE
[ ] OTHER: _____ SPECIFY

3a. Where do you usually park?
[ ] ON STREET
[ ] CITY PARKING STRUCTURE
[ ] LOT BEHIND BUILDING
[ ] CITY LOT NEXT TO LIBRARY
[ ] ELSEWHERE _____ SPECIFY
[ ] NEVER DRIVE MYSELF → GO TO Q.4

3b. Have you ever had problems with parking?
[ ] YES [ ] NO → GO TO Q.3d

3c. What kind of problems? _____

3d. Compared to where you parked before you worked in the Federal Building, is your current parking:
[ ] MORE CONVENIENTLY LOCATED
[ ] LESS CONVENIENTLY LOCATED
[ ] ABOUT THE SAME
[ ] DID NOT DRIVE BEFORE; WASN'T EMPLOYED

4. Before you began working in this building, how did you usually get to and from work?
[ ] OWN CAR
[ ] AGENCY CAR
[ ] SHARE RIDE OR CAR POOL
[ ] BUS
[ ] WALK
[ ] BICYCLE
[ ] OTHER: _____ SPECIFY
[ ] WASN'T EMPLOYED

**3**

5. Since you started working in this building, are you more likely than before to:

a. EAT LUNCH IN A RESTAURANT [ ] YES [ ] NO
b. GO FOR WALK AT LUNCH TIME [ ] YES [ ] NO
c. SHOP DOWNTOWN OR IN THE CAMPUS AREA [ ] YES [ ] NO
d. MEET FRIENDS FOR LUNCH [ ] YES [ ] NO
e. USE THE PUBLIC LIBRARY [ ] YES [ ] NO
f. USE DOWNTOWN RECREATIONAL FACILITIES (Y.M.C.A., ETC.) [ ] YES [ ] NO
g. CONDUCT PERSONAL BUSINESS DOWNTOWN (BANKING, ETC.) [ ] YES [ ] NO

6. How would you rate the location of the Federal Building as a place to work?
[ ] EXCELLENT
[ ] PRETTY GOOD
[ ] FAIR
[ ] POOR

7. Overall, compared to where you worked before, is the location of the Federal Building:
[ ] BETTER    7a. How is it better? _____
[ ] WORSE    7b. How is it worse? _____
[ ] SAME
[ ] WASN'T EMPLOYED

**4**

8. Here are some words used to describe public buildings. Please rate each of the following by placing an X in the box that best describes your feelings about the Federal Building. For example, if you think the building is "attractive" put an X next to the word "attractive." If you think it is "unattractive" put an X right next to the word "unattractive," and if you think it is somewhere in between, please put an X where you think it belongs.

| | | | | | | |
|---|---|---|---|---|---|---|
| ATTRACTIVE | ☐ | ☐ | ☐ | ☐ | ☐ | UNATTRACTIVE |
| WELL KEPT UP INTERIORS | ☐ | ☐ | ☐ | ☐ | ☐ | POORLY KEPT UP INTERIORS |
| WELL KEPT UP ON OUTSIDE | ☐ | ☐ | ☐ | ☐ | ☐ | POORLY KEPT UP ON OUTSIDE |
| POOR ARCHITECTURAL QUALITY | ☐ | ☐ | ☐ | ☐ | ☐ | GOOD ARCHITECTURAL QUALITY |
| DIFFICULT TO FIND WAY AROUND | ☐ | ☐ | ☐ | ☐ | ☐ | EASY TO FIND WAY AROUND |
| UNPLEASANT | ☐ | ☐ | ☐ | ☐ | ☐ | PLEASANT |
| CONVENIENTLY LOCATED TOILETS | ☐ | ☐ | ☐ | ☐ | ☐ | INCONVENIENTLY LOCATED TOILETS |
| ATTRACTIVE INDOOR SIGNS | ☐ | ☐ | ☐ | ☐ | ☐ | UNATTRACTIVE INDOOR SIGNS |
| GOOD OVERALL DESIGN | ☐ | ☐ | ☐ | ☐ | ☐ | POOR OVERALL DESIGN |
| POOR SECURITY | ☐ | ☐ | ☐ | ☐ | ☐ | EXCELLENT SECURITY |
| STIMULATING SPACES | ☐ | ☐ | ☐ | ☐ | ☐ | UNSTIMULATING SPACES |

9. During the past month, how many times have you:

| | NONE | 1-2 TIMES | 3-5 TIMES | 5-10 TIMES | MORE OFTEN |
|---|---|---|---|---|---|
| a. BEEN TO THE BUILDING'S CONFERENCE ROOM | ☐ | ☐ | ☐ | ☐ | ☐ |
| b. USED THE SNACK BAR/CANDY SHOP | ☐ | ☐ | ☐ | ☐ | ☐ |
| c. SAT IN THE LOUNGE OUTSIDE SNACK BAR | ☐ | ☐ | ☐ | ☐ | ☐ |
| d. BOUGHT STAMPS/MAILED THINGS IN POST OFFICE | ☐ | ☐ | ☐ | ☐ | ☐ |
| e. BEEN IN ANOTHER AGENCY IN BUILDING BESIDES POST OFFICE | ☐ | ☐ | ☐ | ☐ | ☐ |
| f. ASKED ASSISTANCE FROM THE SECURITY GUARD | ☐ | ☐ | ☐ | ☐ | ☐ |

10. Overall, how would you rate the building as a place to work?
☐ EXCELLENT
☐ PRETTY GOOD
☐ FAIR
☐ POOR

**5**

These questions deal with the overall space available to your agency—that is, the offices and other work spaces assigned to your organization. Please rate each of the following:

11. The way offices and other work spaces are arranged in terms of making it easier for employees to get their jobs done well:
☐ EXCELLENT
☐ PRETTY GOOD
☐ FAIR
☐ POOR

12. The way the overall space looks:
☐ EXCELLENT
☐ PRETTY GOOD
☐ FAIR
☐ POOR

POST OFFICE EMPLOYEES:
PLEASE SKIP TO PAGE 7

**7**

16. Sometimes the arrangement of offices and work stations can be distracting to people working in a public building. Please indicate how bothersome each of the following is to your work at the Federal Building.

| | NOT AT ALL BOTHERSOME | NOT VERY BOTHERSOME | FAIRLY BOTHERSOME | VERY BOTHERSOME |
|---|---|---|---|---|
| **NOISE** | | | | |
| a. RINGING TELEPHONES IN MY OWN AGENCY | ☐ | ☐ | ☐ | ☐ |
| b. RINGING TELEPHONES IN OTHER AGENCIES | ☐ | ☐ | ☐ | ☐ |
| c. NOISE FROM OTHER EQUIPMENT IN MY OWN AGENCY | ☐ | ☐ | ☐ | ☐ |
| d. NOISE FROM EQUIPMENT IN OTHER AGENCIES | ☐ | ☐ | ☐ | ☐ |
| e. CONVERSATIONS OF OTHERS IN MY AGENCY | ☐ | ☐ | ☐ | ☐ |
| f. CONVERSATIONS OF OTHERS FROM OTHER AGENCIES | ☐ | ☐ | ☐ | ☐ |
| g. NOISE FROM PUBLIC LOBBY/CORRIDORS | ☐ | ☐ | ☐ | ☐ |
| h. NOISE FROM VENTILATING SYSTEM | ☐ | ☐ | ☐ | ☐ |
| i. NOISE FROM STREET OR PARKING LOT | ☐ | ☐ | ☐ | ☐ |
| **LIGHTING** | | | | |
| j. GLARE FROM NATURAL SUNLIGHT | ☐ | ☐ | ☐ | ☐ |
| k. GLARE FROM CEILING LIGHTS | ☐ | ☐ | ☐ | ☐ |
| **HEATING AND VENTILATING** | | | | |
| l. TOO HOT IN SUMMER | ☐ | ☐ | ☐ | ☐ |
| m. TOO COLD IN SUMMER | ☐ | ☐ | ☐ | ☐ |
| n. TOO COLD IN WINTER | ☐ | ☐ | ☐ | ☐ |
| o. TOO HOT IN WINTER | ☐ | ☐ | ☐ | ☐ |
| p. DRAFTS | ☐ | ☐ | ☐ | ☐ |
| q. HEAT FROM NATURAL SUNLIGHT | ☐ | ☐ | ☐ | ☐ |
| r. STUFFY AIR | ☐ | ☐ | ☐ | ☐ |
| **OTHER DISTRACTIONS** | | | | |
| s. PEOPLE WALKING AROUND | ☐ | ☐ | ☐ | ☐ |
| t. FREQUENT REARRANGING OF FURNITURE | ☐ | ☐ | ☐ | ☐ |
| u. FREQUENT REARRANGING OF LIGHTING FIXTURES | ☐ | ☐ | ☐ | ☐ |

**6**

A  B  C  D  E

13. Which type of work area or office arrangement shown above comes closest to the place in which you now work?

☐ TYPE A  ☐ TYPE B  ☐ TYPE C  ☐ TYPE D  ☐ TYPE E

14. Which type comes closest to the place you worked in before coming to the Federal Building?

☐ TYPE A  ☐ TYPE B  ☐ TYPE C  ☐ TYPE D  ☐ TYPE E  ☐ NONE, THIS IS MY FIRST JOB.

15. Which type of office arrangement comes closest to what you would like to work in?

☐ TYPE A  ☐ TYPE B  ☐ TYPE C  ☐ TYPE D  ☐ TYPE E

**8**

17. On an average working day, about how much of your time is spent at your desk or work station?

☐ ALL OR 100 PER CENT
☐ 76 - 99 PER CENT
☐ 51 - 75 PER CENT
☐ 26 - 50 PER CENT
☐ 1 - 25 PER CENT
☐ NONE

18. On an average working day, how often does someone from outside the building come to see you on business?

☐ NEVER
☐ 1 - 2 TIMES
☐ 3 - 4 TIMES
☐ 5 - 10 TIMES
☐ MORE THAN 10 TIMES

19. On an average working day, how many times do you meet with fellow workers at your desk/work station to discuss or perform work?

☐ NEVER
☐ 1 - 2 TIMES
☐ 3 - 4 TIMES
☐ 5 - 10 TIMES
☐ MORE THAN 10 TIMES

20. On an average working day, about how much of the time is spent talking on the telephone?

☐ 76 - 100 PER CENT
☐ 51 - 75 PER CENT
☐ 26 - 50 PER CENT
☐ 11 - 25 PER CENT
☐ 1 - 10 PER CENT
☐ NONE; WORK DOESN'T REQUIRE PHONE CONVERSATIONS

**9**

21. On an average working day, about how many phone conversations do you have?

☐ NONE
☐ 1 - 2
☐ 3 - 4
☐ 5 - 10
☐ MORE THAN 10

22. Do you share a desk or work station with someone else?

☐ YES ☐ NO

23. Please rate your personal work station on each of these characteristics:

| | EXCELLENT | GOOD | FAIR | POOR |
|---|---|---|---|---|
| a. AMOUNT OF SPACE AVAILABLE TO YOU | ☐ | ☐ | ☐ | ☐ |
| b. MATERIALS USED FOR DESKS, TABLES AND CHAIRS | ☐ | ☐ | ☐ | ☐ |
| c. LIGHTING FOR THE WORK YOU DO | ☐ | ☐ | ☐ | ☐ |
| d. LOCATION OF CEILING LIGHTS IN RELATION TO WORK AREA | ☐ | ☐ | ☐ | ☐ |
| e. COLOR OF WALLS AND PARTITIONS | ☐ | ☐ | ☐ | ☐ |
| f. AMOUNT OF SPACE FOR STORING THINGS | ☐ | ☐ | ☐ | ☐ |
| g. ATTRACTIVENESS | ☐ | ☐ | ☐ | ☐ |
| h. CONVERSATIONAL PRIVACY | ☐ | ☐ | ☐ | ☐ |
| i. TYPE OF FLOOR COVERING | ☐ | ☐ | ☐ | ☐ |
| j. YOUR VIEW OUTSIDE | ☐ | ☐ | ☐ | ☐ |
| k. ACCESS TO OTHER PEOPLE YOU HAVE TO WORK WITH | ☐ | ☐ | ☐ | ☐ |
| l. WALL AREA FOR HANGING THINGS (E.G., PICTURES) | ☐ | ☐ | ☐ | ☐ |
| m. STYLE OF YOUR FURNITURE | ☐ | ☐ | ☐ | ☐ |
| n. NUMBER OF ELECTRICAL OUTLETS | ☐ | ☐ | ☐ | ☐ |
| o. LOCATION OF ELECTRICAL OUTLETS | ☐ | ☐ | ☐ | ☐ |
| p. VISUAL PRIVACY | ☐ | ☐ | ☐ | ☐ |
| q. AMOUNT OF SURFACE AREA FOR WORK | ☐ | ☐ | ☐ | ☐ |
| r. COMFORT OF YOUR CHAIR | ☐ | ☐ | ☐ | ☐ |
| s. OVERALL AESTHETIC QUALITY | ☐ | ☐ | ☐ | ☐ |
| t. VENTILATION AND AIR CIRCULATION | ☐ | ☐ | ☐ | ☐ |
| u. HEATING | ☐ | ☐ | ☐ | ☐ |
| v. AIR QUALITY | ☐ | ☐ | ☐ | ☐ |

**10**

24. Compared to where you worked before coming to the Federal Building, is your present work station:

- [ ] BETTER — 24a. How is it better? _____
- [ ] WORSE — 24b. How is it worse? _____
- [ ] SAME
- [ ] WASN'T EMPLOYED

25. Here are some statements about peoples' jobs. Please indicate how true each is in your job.

| | VERY TRUE | SOMEWHAT TRUE | NOT VERY TRUE | NOT AT ALL TRUE |
|---|---|---|---|---|
| TRAVEL TO AND FROM WORK IS CONVENIENT | [ ] | [ ] | [ ] | [ ] |
| THE WORK IS INTERESTING | [ ] | [ ] | [ ] | [ ] |
| WHENEVER I TALK TO CO-WORKERS, OTHERS CAN HEAR OUR CONVERSATIONS | [ ] | [ ] | [ ] | [ ] |
| I DO AS MUCH WORK AS I REASONABLY CAN | [ ] | [ ] | [ ] | [ ] |
| I AM GIVEN ALOT OF CHANCES TO MAKE FRIENDS | [ ] | [ ] | [ ] | [ ] |
| I HAVE THE OPPORTUNITY TO DEVELOP MY OWN SPECIAL ABILITIES | [ ] | [ ] | [ ] | [ ] |
| WHENEVER I TALK ON THE TELEPHONE OTHER: AROUND ME CAN HEAR MY CONVERSATIONS | [ ] | [ ] | [ ] | [ ] |
| THE PEOPLE IN MY AGENCY DO AS MUCH WORK AS THEY REASONABLY CAN | [ ] | [ ] | [ ] | [ ] |
| I HAVE ACCESS TO THE EQUIPMENT AND MATERIAL I NEED TO GET THE JOB DONE WELL | [ ] | [ ] | [ ] | [ ] |
| THE PHYSICAL SURROUNDINGS ARE PLEASANT | [ ] | [ ] | [ ] | [ ] |
| COMPARED TO WHERE I WORKED BEFORE COMING TO THIS BUILDING, I DO MORE WORK NOW | [ ] | [ ] | [ ] | [ ] |
| MY WORK SURFACE, STORAGE SPACE, CHAIR AND OTHER FURNITURE ARE WHAT I NEED TO GET THE JOB DONE WELL | [ ] | [ ] | [ ] | [ ] |

26. Overall, how satisfied are you with your work station:

- [ ] VERY SATISFIED
- [ ] FAIRLY SATISFIED
- [ ] NOT VERY SATISFIED
- [ ] NOT AT ALL SATISFIED

**11**

27. Which of the following best describes your job?

- [ ] MANAGER-SUPERVISOR
- [ ] PROFESSIONAL-TECHNICAL
- [ ] CLERICAL-SECRETARIAL
- [ ] POSTAL CARRIER
- [ ] MILITARY RECRUITMENT

28. How many days during the week are you usually at work in the Federal Building?

- [ ] 2 DAYS OR LESS PER WEEK
- [ ] 3 - 4 DAYS PER WEEK
- [ ] 5 DAYS PER WEEK
- [ ] MORE THAN 5 DAYS PER WEEK

29. During your average working day, how many times do you leave the building in connection with your work?

- [ ] NONE, NEVER LEAVE BUILDING → GO TO Q.30
- [ ] 1 - 2 TIMES
- [ ] 3 - 4 TIMES
- [ ] 5 OR MORE TIMES

29a. Do you usually drive when you leave the building?
[ ] YES   [ ] NO

30. About how long does it take you to get to work?

- [ ] LESS THAN 15 MINUTES
- [ ] 15 - 29 MINUTES
- [ ] 30 - 44 MINUTES
- [ ] 45 - 59 MINUTES
- [ ] ONE HOUR OR MORE

31. Are you:
[ ] FEMALE   [ ] MALE

This completes the questionnaire. Thank you for your cooperation. If you have any additional comments about the building, please feel free to write them down on the back of this page.

Architectural Research Laboratory
College of Architecture and Urban Planning
The University of Michigan

INTERVIEW NO._____

DATE:_____

TIME IS NOW:_____

## OUTSIDE USERS QUESTIONNAIRE

Excuse me, do you work in this building?  [5.] NO  [1.] YES——►Thank you.  END CONTACT

I'm _____ from the University of Michigan's Architectural Research
Laboratory and we're working on a study which deals with this building.  Would you mind
answering a few questions - it should only take a few minutes.  PAUSE

Good.  Before we start, I want to assure you that the interview is completely
voluntary.  If we should come to a question which you don't want to answer, just
let me know and we'll go to the next question.

1.  Is this the first time you've been to this building?

   [5.] NO    [1.] YES ——GO TO Q.6

2.  About how many times during the past month have you been here?

   _____ NUMBER OF TIMES

3.  Have you (ever) been to the post office here (before)?

   [1.] YES   [5.] NO

4.  How about other agencies here:  Have you ever been to any of them?

   [1.] YES   [5.] NO

5.  And when you come to the building, do you usually drive, walk, take a bus, or what?

   [1.] DRIVE  [2.] WALK  [3.] BUS  [4.] BICYCLE  [7.] OTHER:_____ SPECIFY

2

6. How did you get here this time (on this trip). Did you drive, walk, take a bus, or what?

1. DRIVE  2. WALK  3. BUS  4. BICYCLE  7. OTHER: _____ SPECIFY

GO TO Q.7

6a. Where did you park? _____

6b. Did you have problems with parking?

1. YES  5. NO → GO TO Q.6d

6c. What kind of problems did you have? _____

6d. In general, how convenient is parking around here. Would you say it is very convenient, fairly convenient, not very convenient, or not at all convenient?

1. NOT CONVENIENT  2. FAIRLY CONVENIENT  4. NOT VERY CONVENIENT  5. NOT AT ALL CONVENIENT

7. Did you (ever) have difficulty in finding your way to the places you had to go to in the building?

1. YES  5. NO

8. Did you (ever) use the:

9. Did you (ever) have difficulty finding the _____?

a. elevators                        5. NO  1. YES → ASK Q.9   a. 1. YES  5. NO
b. stairs                           5. NO  1. YES → ASK Q.9   b. 1. YES  5. NO
c. snack bar or candy counter       5. NO  1. YES → ASK Q.9   c. 1. YES  5. NO
d. restrooms                        5. NO  1. YES → ASK Q.9   d. 1. YES  5. NO
e. information desk                 5. NO  1. YES → ASK Q.9   e. 1. YES  5. NO

3

10. Did you (ever) wander or look around the building, that is just to explore it?

1. YES  5. NO

11. What do you think of the appearance of the inside of the building? Would you say it's very attractive, fairly attractive, not very attractive, or not at all attractive?

1. VERY ATTRACTIVE  2. FAIRLY ATTRACTIVE  4. NOT VERY ATTRACTIVE  5. NOT AT ALL ATTRACTIVE  8. DON'T KNOW

12. And what do you think about the overall appearance of the outside of the building. Is it very attractive, fairly attractive, not very attractive, or not at all attractive?

1. VERY ATTRACTIVE  2. FAIRLY ATTRACTIVE  4. NOT VERY ATTRACTIVE  5. NOT AT ALL ATTRACTIVE  8. DON'T KNOW

13. Is there anything about the building you especially like?

1. YES  5. NO → GO TO Q.14

13a. What do you like about it? _____

14. Is there anything about the building you especially don't like?

1. YES  5. NO → GO TO Q.15

14a. What don't you like about the building? _____

15. How well do you think the building fits into downtown Ann Arbor. Does it fit in very well, fairly well, not very well, or not well at all?

1. VERY WELL  2. FAIRLY WELL  4. NOT VERY WELL  5. NOT WELL AT ALL  8. DON'T KNOW

16. What about the plaza or open area in front of the building? Would you say it's very attractive, fairly attractive, not very attractive, or not at all attractive?

| 1. VERY ATTRACTIVE | 2. FAIRLY ATTRACTIVE | 4. NOT VERY ATTRACTIVE | 5. NOT AT ALL ATTRACTIVE | 8. DON'T KNOW |

17. Have you ever used the plaza?

1. YES    5. NO →GO TO Q.18

17a. How did you use it? _____

18. Those are all the questions I have about the Federal Building. Now I'd like to ask you just a few more questions of a general nature. Do you work in Ann Arbor?

1. YES    5. NO→GO TO Q.19    8. NOT CURRENTLY EMPLOYED ──TO TO Q.19

18a. And what part of town do you work in? _____

18b. (IF DOWNTOWN ASK:) What are the nearest cross streets to where you work? _____

19. Are you a student at the University of Michigan?

1. YES    5. NO

20. About how many times during the past month have you been to downtown Ann Arbor to shop, eat out or conduct personal business? By downtown I mean the area west of State Street, including Main Street. _____ NUMBER OF TIMES

21. Do you live in the city of Ann Arbor or one of the surrounding townships?

1. YES    5. NO→GO TO END OF INTERVIEW

21a. About how long have you lived in the Ann Arbor area? _____ YEARS _____ MONTHS

That completes the interview, thanks for your cooperation.

TIME IS NOW: _____

LENGTH OF INTERVIEW: _____ MINUTES

---

INTERVIEWER COMMENTS:

22. RESPONDENT WAS:

a.   1. MALE    2. FEMALE

b.   1. WHITE    2. BLACK    7. OTHER

c.   1. VERY COOPERATIVE    2. FAIRLY COOPERATIVE    4. NOT VERY COOPERATIVE

d.   1. ABOVE AVERAGE INTELLIGENCE    2. AVERAGE INTELLIGENCE    4. BELOW AVERAGE INTELLIGENCE

e.   1. NICELY/NEATLY DRESSED    2. AVERAGE DRESSED    4. POORLY/SLOPPILY DRESSED

f.   1. UNDER 25 YEARS OF AGE    2. 25 – 29 YEARS OF AGE    3. 50 – 64 YEARS OF AGE    4. 65 YEARS OF AGE OR OLDER

23. WEATHER WAS:

a.   1. CLEAR    2. RAINY    3. SNOWING    4. OVERCAST

b.   1. MILD (OVER 32°)    2. COLD (20° – 32°)    3. VERY COLD (LESS THAN 20°)

24. INTERVIEW TAKEN AT:

a.   1. MAIN ENTRANCE    2. NEAR POST OFFICE ENTRANCE

   3. FAR POST OFFICE ENTRANCE    4. REAR ENTRANCE

OTHER COMMENTS:

Architectural Research Laboratory
College of Architecture and Urban Planning
The University of Michigan

INTERVIEW NO. _____

## COMMUNITY - AT - LARGE  QUESTIONNAIRE

TIME IS NOW: _____

1. First of all, have you ever been in the Ann Arbor Federal Building?

[1.] YES    [5.] NO    [8.] NOT SURE

1a. Do you know where the Federal Building is?

[1.] YES    [5.] NO    [8.] NOT SURE

GO TO Q.1c

1b. Could you tell me where it is?

_____

_____

INTERVIEWER:

☐ R KNOWS BUILDING IS DOWNTOWN ──→ GO TO Q.2
☐ R DOES NOT KNOW BUILDING IS DOWNTOWN.

1c. Are you familiar with the light-brown tiled building downtown with the post office in it, the one on Liberty Street?

[1.] YES    [5.] NO ──→ GO TO Q.20

1d. Well that's Ann Arbor's Federal Building. Have you ever been in the building?

[1.] YES    [5.] NO ──→ GO TO Q.11

2. About how many times during the past month have you been there?

_____  NUMBER OF TIMES

3. Have you ever been to the post office in the building?

[1.] YES    [5.] NO

4. How about other agencies?  Have you ever been to any of them?

[1.] YES    [5.] NO

---

Architectural Research Laboratory
College of Architecture and Urban Planning
The University of Michigan

COVER SHEET NO. _____
INTERVIEW NO. _____

## COMMUNITY - AT - LARGE  QUESTIONNAIRE

### COVER  SHEET

PHONE NUMBER _____    DIRECTORY PAGE NO. _____

ADDRESS _____    LINE NO. _____

1. Hello, my name is _____.  I'm calling from the University of Michigan here in Ann Arbor.  We are currently working on a study for the Architectural Research Laboratory.  This study deals with Ann Arbor's Federal Office Building.  I'd like to ask a few questions of someone in your household who is at least 18 years of age. Are you 18 or older?

[1.] YES

[5.] NO

1a. Is there someone over 18 at home whom I can talk with?

[1.] YES    [5.] NO ──→ GO TO Q.1c

1b. May I speak with that person?

[1.] YES    [5.] NO ──→ GO TO Q.1c

INTERVIEWER: REPEAT INTRODUCTION FOR NEW RESPONDENT.

1c. When would be a good time to call back?

Thank you for your help.  Good-bye.

INTERVIEWER: RECORD CALL INFORMATION BELOW.

Good.  Before we start, I want to assure you that the interview is completely voluntary. If we should come to a question which you don't want to answer, just let me know and we'll go to the next question.

| Call Number | 01 | 02 | 03 | 04 |
|---|---|---|---|---|
| Date | | | | |
| Time | | | | |
| Result | | | | |
| I'er Initials | | | | |

R = Refusal
NA = No answer
CB = Call back
IT = Interview Taken

EXPLAIN REFUSALS AT END OF PAGE 6.

2

5. And when you go to the Federal Building, do you usually drive, walk, take a bus, or what?

[1.] DRIVE   [2.] WALK   [3.] BUS   [4.] BICYCLE   [7.] OTHER: _____ SPECIFY
                                                              → GO TO Q.6

5a. Where do you usually park?

5b. Did you ever have difficulty with the parking there?

[1.] YES   [5.] NO → GO TO Q.5d

5c. What kind of problems did you have?

5d. In general, how convenient is the parking? Would you say it's very convenient, fairly convenient, not very convenient or not at all convenient?

[1.] VERY CONVENIENT   [2.] FAIRLY CONVENIENT   [4.] NOT VERY CONVENIENT   [5.] NOT AT ALL CONVENIENT

6. In your visit(s) to the building, did you ever have difficulty in finding your way to the offices or other places you've had to go to?

[1.] YES   [5.] NO

7. Did you ever use the:          8. Did you (ever) have difficulty finding the _____?

| | 7. | | 8. | |
|---|---|---|---|---|
| a. elevators | [5.] NO | [1.] YES → ASK Q.8→a. | [1.] YES | [5.] NO |
| b. stairs | [5.] NO | [1.] YES → ASK Q.8→b. | [1.] YES | [5.] NO |
| c. snack bar or candy counter | [5.] NO | [1.] YES → ASK Q.8→c. | [1.] YES | [5.] NO |
| d. restrooms | [5.] NO | [1.] YES → ASK Q.8→d. | [1.] YES | [5.] NO |
| e. information desk | [5.] NO | [1.] YES → ASK Q.8→e. | [1.] YES | [5.] NO |

9. Did you ever wander or look around the building, that is just to explore it?

[1.] YES   [5.] NO

3

10. What do you think of the appearance of the inside of the building. Would you say it's very attractive, fairly attractive, not very attractive or not at all attractive?

[1.] VERY ATTRACTIVE   [2.] FAIRLY ATTRACTIVE   [4.] NOT VERY ATTRACTIVE   [5.] NOT AT ALL ATTRACTIVE   [8.] DON'T KNOW

11. And what do you think of the overall appearance of the outside of the building. Would you say it is very attractive, fairly attractive, not very attractive or not at all attractive?

[1.] VERY ATTRACTIVE   [2.] FAIRLY ATTRACTIVE   [4.] NOT VERY ATTRACTIVE   [5.] NOT AT ALL ATTRACTIVE   [8.] DON'T KNOW

12. Is there anything about the building you especially like?

[1.] YES   [5.] NO → GO TO Q.13

12a. What do you like about it? _____

13. Is there anything about the building you especially don't like?

[1.] YES   [5.] NO → GO TO Q.15

14. What don't you like about the building? _____

15. How well do you think the building fits into downtown Ann Arbor. Does it fit in very well, fairly well, not very well, or not well at all?

[1.] VERY WELL   [2.] FAIRLY WELL   [5.] NOT VERY WELL   [5.] NOT WELL AT ALL   [8.] DON'T KNOW

16. What about the plaza or open area in front of the building. Would you say it's very attractive, fairly attractive, not very attractive or not at all attractive?

[1.] VERY ATTRACTIVE   [2.] FAIRLY ATTRACTIVE   [4.] NOT VERY ATTRACTIVE   [5.] NOT AT ALL ATTRACTIVE   [8.] DON'T KNOW

5

23. Do you live in the city of Ann Arbor or one of the surrounding townships?

[1.] YES  [5.] NO ⟶ GO TO END OF INTERVIEW

23a. About how long have you lived in the Ann Arbor area? _____ YEARS _____ MONTHS

That completes the interview. Thank you for your cooperation.

TIME IS NOW: _____

LENGTH OF INTERVIEW: _____ MINUTES

---

4

17. Have you ever used the plaza?

[1.] YES  [5.] NO ⟶ GO TO Q.18

17a. How did you use it? _____

18. How many people would you guess work in the Federal Building? (IF RANGE MORE THAN 100 ASK FOR SPECIFIC NUMBER)

_____ PEOPLE  999. [R HAS NO IDEA AND WON'T GUESS]

GO TO Q.19

18a. Would you guess that more, less or about the same number of people work in the Federal Building than in City Hall?

[1.] MORE  [5.] LESS  [3.] ABOUT THE SAME  [8.] DON'T KNOW

GO TO Q.20

19. And for a comparison, how many would you guess work in the Ann Arbor City Hall on Huron Street?

_____ PEOPLE  999. [R HAS NO IDEA AND WON'T GUESS]

20. Those are all of the questions I have about the Federal Building. Now I'd like to ask you just a few more questions of a general nature. Do you work in Ann Arbor?

[1.] YES  [5.] NO ⟶ GO TO Q.21  [8.] NOT CURRENTLY EMPLOYED ⟶ GO TO Q.21

20a. And what part of town do you work in? _____

20b. (IF DOWNTOWN ASK:) What are the nearest cross streets to where you work? _____

21. Are you a student at the University of Michigan?

[1.] YES  [5.] NO

22. About how many times during the past month have you been to downtown Ann Arbor for shopping, eating out, or conducting personal business? By downtown I mean the area west of State Street, including Main Street.

_____ NUMBER OF TIMES

6

INTERVIEWER COMMENTS:

24. RESPONDENT WAS:

a.  [1.] MALE      [2.] FEMALE

b.  [1.] VERY          [2.] FAIRLY          [4.] NOT VERY
        COOPERATIVE          COOPERATIVE          COOPERATIVE

c.  [1.] VERY INTERESTED    [2.] FAIRLY          [4.] NOT VERY
        IN STUDY                INTERESTED            INTERESTED

d.  [1.] ABOVE AVERAGE      [2.] AVERAGE         [4.] BELOW AVERAGE
        INTELLIGENCE            INTELLIGENCE          INTELLIGENCE

OTHER COMMENTS:

REASON FOR REFUSAL:

# Appendix C

## Data Collection Instruments
## for the Environmental Measures

Architectural Research Laboratory
College of Architecture and Urban Planning
The University of Michigan

SPACE ID NO. _____
AGENCY _____
RECORDER _____

## ENVIRONMENTAL MEASURES - INDIRECT

1. TYPE OF WORKSPACE:

   ☐ Conventional work station: Own
   ☐ Conventional work station: Shared
   ☐ Open work station with partitions: Own
   ☐ Open work station with partitions: Shared
   ☐ Open work station without partitions
   ☐ Other:
   (SPECIFY)

2. AMOUNT OF WORKSPACE:

   _____ Square Feet

   Area measured in square feet contiguous to main work surface of worker. Include area occupied by desks, chairs, tables, and other furniture/equipment associated with worker. In enclosed spaces with one worker, area is defined by enclosing walls or partitions.

   In enclosed or open spaces with multiple workers, area is limited to that containing individual worker's furniture/equipment and space 3'-0" beyond it, unless that space infringes on space of neighboring worker, in which case space is defined by half the distance to nearest piece of furniture/equipment.

3. DENSITY OF WORKSPACE:

   _____ Persons/400 Square Feet

   Number of people who work within a 400 square foot area whose centroid is in the center of the main work surface (USE TEMPLATE)

4. STRAIGHT-LINE DISTANCE TO NEAREST USABLE AGENCY ENTRANCE:

   ☐ Obstructed straight-line
   _____ Feet (straight-line distance)

   Straight-line distance measured in feet between nearest usable entrance of agency and center of desk or main work surface. If the straight-line between the entrance and the work surface is obstructed by a floor to ceiling partition, no distance is recorded.

5. FUNCTIONAL DISTANCE TO NEAREST USABLE AGENCY ENTRANCE:

   _____ Feet (functional distance)

   Functional distance measured in linear feet between nearest usable entrance and center of desk or main work surface.

6. DISTANCE TO NEAREST WINDOW:

   ☐ No window in agency
   ☐ Agency has window but worker is in enclosed space with no glass partition
   ☐ Agency has window and worker is in an open space or an enclosed space with a glass partition
   _____ Feet (straight-line distance)

   Distance measured in linear feet between center of worker's desk or main work surface and nearest vertical glazed surface separating inside from outside of building.(use straight-line distance)

7. DISTANCE TO AGENCY'S COFFEE POT:

   ☐ No coffee pot in agency
   _____ Feet (functional distance)

   Distance measured in linear feet between center of worker's desk or main work station and nearest coffee station within agency. (use functional distance)

8. DISTANCE TO LIGHT WELL ABOVE:

   ☐ No light well above agency
   ☐ Light well above agency but work station is located in enclosed space with no glass partition
   ☐ Light well above agency and work station is located in an open space or an enclosed space with a glass partition
   _____ Feet (straight-line distance)

   Distance measured at a right angle from the nearest edge of the light well above to the center of the worker's desk or main work station.

9. DISTANCE TO LIGHT WELL BELOW:

   ☐ No light well below agency
   ☐ Light well below agency but work station is located in enclosed space with no glass partition
   ☐ Light well below agency and work station is located in an open space or an enclosed space with a glass partition
   _____ Feet (straight-line distance)

   Distance measured at a right angle from the nearest edge of the light well below to the center of the worker's desk or main work station.

-3-

8. Agency/Neighbors' Decor (Objects)

☐ None without options
☐ None with options
☐ 1 or 2 objects
☐ 3 to 4 objects
☐ 5 or more objects

**Objects include pictures, plants, ashtrays, or other items provided by the agency or other workers and are positioned so they can be seen from the work station chair.**

9. Electrical Outlets:

☐ Work station is directly over or touching electrical outlet
☐ Work station is supplied electrical connection with an extension cord
☐ Work station does not have electrical service

10. Communication Outlets: (Telephones, Teletype, Intercom)

☐ Work station is directly over or touching electrical outlet
☐ Work station is supplied electrical connection with an extension cord
☐ Work station does not have electrical service

11. Chair:

☐ Contour chair with full padding, armrests, swivel and rollers
☐ Standard government issue chair, padded, swivel, rollers, without armrests
☐ Standard government issue chair, padded, without swivel or rollers
☐ Unpadded chair
☐ Stool
☐ Other: _____ (SPECIFY)

-4-

12. WORK TASK:

☐ Interviewing customers from a desk
☐ Typing and steno
☐ Filing papers, writing, or desk calculations
☐ Drafting
☐ Sorting mail
☐ Technical or high intensity bench work
☐ Computer terminal access
☐ Postal customer services
☐ Other customer services
☐ Reception from a desk
☐ Reception and typing from a desk
☐ Other: _____ SPECIFY

**Check all tasks that are apparent for each work station.**

13. ZONAL NOISE:

NC Reading: _____
Hz Reading: _____
dBA Reading: _____

NC is Noise Criteria measured on standard octave bands.

Hz is Hertz level of highest NC penetration.

dBA is weighted average ("A" method) of noise level.

14. ZONAL TEMPERATURE:

1st Reading: ___° Date: _____ Time: _____
2nd Reading: ___° Date: _____ Time: _____

15. ZONAL HUMIDITY:

1st Reading: ___ rh Date: _____ Time: _____
2nd Reading: ___ rh Date: _____ Time: _____

Architectural Research Laboratory
College of Architecture and Urban Planning
The University of Michigan

SPACE ID NO. ____
AGENCY ____
RECORDER ____

## ENVIRONMENTAL MEASURES - DIRECT

1. TYPE OF WORK STATION:

   ☐ Conventional work station: Own
   ☐ Conventional work station: Shared
   ☐ Open work station with partitions: Own
   ☐ Open work station with partitions: Shared
   ☐ Open work station without partitions
   ☐ Other: _____
                (SPECIFY)

2. WORKERS RELATION TO STATION OVER TIME:

   ☐ Worker was sitting at work station 3 months ago.
   ☐ Worker was not sitting at work station 3 months ago.
   ☐ Other: _____
                (SPECIFY)

3. LIGHT ENVIRONMENTAL ZONE:

   ☐ Closed office with no external natural light
   ☐ Closed office with external natural light
   ☐ Open office, more than 10 feet from light well
     or more than 20 feet from window
   ☐ Open office and within 20 feet of window
   ☐ Open office and within 10 feet of light well

-2-

4. WORK STATION LIGHT LEVEL:

   OUTSIDE CONDITION:

   1st Reading: ____ fc      ☐ Sunny  ☐ Hazy      Date: ____   Time: ____
   2nd Reading: ____ fc      ☐ Sunny  ☐ Hazy      Date: ____   Time: ____

5. WORK STATION GLARE CONDITION:

   ☐ Within 20 feet of north window and facing south
   ☐ Within 10 feet of light well and facing away from wall
   ☐ Within 20 feet of south, east, or west window
   ☐ More than 20 feet from window or more than 10 feet from light well
   ☐ Within 20 feet of north window and facing north
   ☐ Within 20 feet of north window and facing east or west
   ☐ Within 10 feet of light well and facing light well

WORK STATION DECOR:

6. Individual Task Lighting

   ☐ None
   ☐ Incandescent
   ☐ Florescent

7. Individual Decor (Objects)

   ☐ None without options
   ☐ None with options
   ☐ 1 or 2 objects
   ☐ 3 to 4 objects
   ☐ 5 or more objects

Objects include pictures, plants, ashtrays, wall hangings or other personal momentos, located within the individual work station and located on the walls, partitions, floor, desk or other work surface. Without options indicates that the worker has no surface or wall space suitable for placing an object or is unable to do so by virtue of agency restrictions.

# Appendix D

## Environmental Zone Definitions

---

### Work Setting

*Industrial.* A classification of work setting used solely for the work environment found in the Post Office. It refers to an environment with no interior partitions and work stations separated primarily by functional pieces of mail sorting equipment and moveable sorting carts. The setting is characterized by high ceilings with exposed structural steel, industrial fluorescent lighting, no carpeting and a highly reflective wall treatment.

*Conventional.* A work setting housing seven or fewer employees within a self-contained and lockable room, separated from other individuals or agencies by full height partitions and secure entries. This type of work setting is characterized by small, unified groups of people or individuals that work in areas with full acoustical surface treatment, privacy from other agencies, and well-defined work stations separated from large numbers of other workers or the public.

*Open offices.* Open office work settings contain entire agencies or groups of individuals that work in a large, common space. Although the space has essentially the same physical amenities found in the conventional settings, the open office is distinct from the conventional in that no full height partitions separate the workers either from their own agency co-workers or, in some cases, from adjacent agencies. The open office is characterized by large, integrated work areas (with work stations), separated by moveable or no partitions, and sharing a common entrance and security system.

## Exterior Environment Connection

*Light and view.* A condition defined by the presence of a vertical glazed surface that allows the work station occupant to view the exterior environment and receive natural sunlight. Two types of natural lighting and view are defined as a) a south facing window shared by a conventional work setting group or b) a north facing window wall shared by an open office work station group. It should be noted that no conventional work station occupied by single individuals had either of these connections to the exterior.

*Light only.* In instances where work stations are located more than 20 feet from vertical glazed areas but within 10 feet of an interior skylight or lightwell, the work station is defined as having a light connection only. This connection is characterized by the lack of an exterior view, either because no vertical glazed surfaces are in sight or because partitions or other work stations block an exterior view. The work station is connected to the exterior in the sense that changing light conditions can be detected and a partial view of the sky is afforded.

*Closed.* An environment characterized by the lack of direct contact with the exterior, either through vertical windows or skylights. In some instances, these work stations are located within agencies with no windows or skylights at all or within areas of agencies far removed and visually separated from such connections to the exterior.

# References

Alexander, C.; Ishikawa, S.; and Silverstein, M. *A Pattern Language.* New York: Oxford University Press, 1977.

Andrews, F. M.; Morgan, J. N.; Sonquist, J. A.; and Klem, L. *Multiple Classification Analysis: A Report on a Computer Program for Multiple Regression Using Categorical Predictors, Second Edition.* Ann Arbor: Institute for Social Research, The University of Michigan, 1973.

*Architectural Record.* "The New Building: Those Guiding Principles" (December 1978): 110–111.

———. "Three Federal Buildings: How'd They Get So Good?" (December 1978): 112–121.

Campbell, A.; Converse, P. E.; and Rodgers, W. L. *The Quality of American Life.* New York: Russell Sage Foundation, 1976.

Canter, D.; Kenny, C.; and Rees, K. "A Multivariate Model for Place Evaluation." Department of Psychology, University of Surrey, England, 1980.

Cooper, C. C. *Easter Hill Village: Some Social Implications of Design.* New York: The Free Press, 1975.

Deegan, J. "On the Occurrence of Standardized Regression Coefficients Greater than One." *Educational and Psychological Measurement* (Winter 1978): 873–888.

Friedman, A.; Zimering, C.; and Zube, E. *Environmental Design Research.* New York: Plenum Press, 1978.

Harris, Louis, and Associates, Inc. *The Steelcase National Survey of Office Environments: Do They Work?* Steelcase, Inc., 1978.

Lansing, J. B.; Marans, R. W.; and Zehner, R. B. *Planned Residential Environments.* Ann Arbor: Institute for Social Research, The University of Michigan, 1970.

Marans, R. W. "Evaluation Research and Its Uses by Housing Designers and Managers." Institute for Social Research, The University of Michigan, Ann Arbor, 1979. Mimeograph.

Marans, R. W., and Fly, J. M. *Recreation and the Quality of Urban Life: Recreational Resources, Behaviors, and Evaluations of People in the Detroit Region.* Institute for Social Research, The University of Michigan, Ann Arbor, 1981.

————, and Rodgers, W. "Toward an Understanding of Community Satisfaction." In *Metropolitan American in Contemporary Perspectives,* edited by A. Hawley and V. Rock. New York: Halsted Press, 1975.

Nunnally, J. D. *Psychometric Theory.* New York: McGraw-Hill Book Company, 1967.

*Progressive Architecture.* Editorial by J. M. Dixon (January 1980): 7–8.

Quinn, R. P. "Effectiveness in Work Roles: Employee Responses to Work Environments." Institute for Social Research, The University of Michigan, Ann Arbor, 1977.

————, and Staines, G. L. *The 1977 Quality of Employment Survey.* Institute for Social Research, The University of Michigan, Ann Arbor, 1979.

United Nations. *The Role of Housing in Promoting Social Integration.* New York: Department of Economic and Social Affairs, United Nations, 1978.